Newton's Apple is Now The Fermion

Bottom Up Birth of Cosmos Theory from Geometry
Spaces Within Spaces; Universes Within Universes
Symmetry and Fermion Structures from su(1, 1)
New Fundamental View of Interactions
Superluminal Physics
HubbaHubble Expansion of Universes
A Big Dip in Universe Expansion
Fibonacci & Ramsey Numbers

Stephen Blaha Ph. D.
Blaha Research

Pingree-Hill Publishing

MMXXIII

Copyright © 2023 by Stephen Blaha. All Rights Reserved.

This document is protected under copyright laws and international copyright conventions. No part of this book may be reproduced, stored in a retrieval system, or transmitted by any means in any form, electronic, mechanical, photocopying, recording, or as a rewritten passage(s), or otherwise, without the express prior written permission of Blaha Research.

ISBN: 978-1-7356795-6-3

This document is provided "as is" without a warranty of any kind, either implied or expressed, including, but not limited to, implied warranties of fitness for a particular purpose, merchantability, or non-infringement. This document may contain typographic errors or technical inaccuracies. This book is printed on acid free paper.

Cover: A photograph appears on the cover courtesy of NASA/ESA/Hubble.

Rev. 00/00/01 June 26, 2023

To Margaret

Some Other Books by Stephen Blaha

All the Megaverse! Starships Exploring the Endless Universes of the Cosmos using the Baryonic Force (Blaha Research, Auburn, NH, 2014)

SuperCivilizations: Civilizations as Superorganisms (McMann-Fisher Publishing, Auburn, NH, 2010)

All the Universe! Faster Than Light Tachyon Quark Starships & Particle Accelerators with the LHC as a Prototype Starship Drive Scientific Edition (Pingree-Hill Publishing, Auburn, NH, 2011).

Unification of God Theory and Unified SuperStandard Model THIRD EDITION (Pingree Hill Publishing, Auburn, NH, 2018).

The Exact QED Calculation of the Fine Structure Constant Implies ALL 4D Universes have the Same Physics/Life Prospects (Pingree Hill Publishing, Auburn, NH, 2019).

Integration of General Relativity and Quantum Theory: Octonion Cosmology, GiFT, Creation/Annihilation Spaces CASe, Reduction of Spaces to a Few Fermions and Symmetries in Fundamental Frames (Pingree Hill Publishing, Auburn, NH, 2021).

Passing Through Nature to Eternity ProtoCosmos, HyperCosmos, Unified SuperStandard Theory (Pingree Hill Publishing, Auburn, NH, 2022).

HyperCosmos Fractionation and Fundamental Reference Frame Based Unification: Particle Inner Space Basis of Parton and Dual Resonance Models (Pingree Hill Publishing, Auburn, NH, 2022).

A New UniDimension ProtoCosmos and SuperString F-Theory Relation to the HyperCosmos (Pingree Hill Publishing, Auburn, NH, 2022).

The Cosmic Panorama: ProtoCosmos, HyperCosmos, Unified SuperStandard Theory (UST) Derivation (Pingree Hill Publishing, Auburn, NH, 2022).

Ultimate Origin: ProtoCosmos and HyperCosmos (Pingree Hill Publishing, Auburn, NH, 2022).

A New Completely Geometric SU(8) Cosmos Theory; New PseudoFermion Fields; Fibonacci-like Dimension Arrays; Ramsey Number Approximation (Pingree Hill Publishing, Auburn, NH, 2023).

Available on Amazon.com, bn.com, Amazon.co.uk and other international web sites as well as at better bookstores.

CONTENTS

FIGURES and TABLES

Introduction

This book presents important new aspects of Cosmos Theory. It starts with a comparison of a Top Down approach to Cosmos Theory and a Bottom Up approach to Cosmos Theory.

The bottom up derivation begins with the relation of Minkowski space to the Spin Group su(2, 2). It leads directly to the basis of Cosmos Theory in fermion spin states of spaces of various dimensions. Since the number of spin states for any space-time is a power of 2 we are led to the form of the 10 HyperCosmos and 10 Second Kind HyperCosmos spaces that similarly use powers of two.

We chose to create a UniDimension ProtoCosmos that generates the Geometric spectrum of these spaces in analogy to the Hydrogen atom. We generalize this atom to have a fermion wave function factor in analogy to angular momentum in the Hydrogen atom.

We define s new type of fermion wave function for two purposes: one purpose was to define a universe for the ProtoCosmos; the other purpose was to define a universe for each HyperCosmos and Second Kind HyperCosmos space. This requirement led to our development of Independent PseudoFermion (IPF) wave functions – a generalization of the PseudoFermion wave functions that we introduced in the preceding book.

An IPF wave function has independent dimension arrays for the parent and child spaces. Thus the child space has a separate dimension array with which it can create a universe of symmetries and particles.

These developments led to the following topics in the book:

1. A su(1, 1) derivation of the structure of symmetries and particles in the Fundamental Reference Frames (FRFs).

2. A detailed analysis of FRF contents for each type of space and the General Relativistic transformations to the dimension arrays of universes.

3. A complete description of the Unified SuperStandard Theory (UST) for our universe.

4. A detailed description of the set of UST interactions including The Generation Group, Layer Group, Connection Groups, and the Strong and ElectroWeak interactions. Also a 4 ×4 Cabibbo-Kobayashi-Maskawa Matrix discussion.

5. A description of the generation of hierarchies of universes: child, sibling, grandchildren and so on. The possibility of a network of universes is also treated.

6. There is a possibility of a zero space-time dimension Second Kind HyperCosmos space supporting a sea of miniverses within our universe. These miniverses could comprise Dark Matter. They may be anyon "particles." They would only have gravity interactions – agreeing with current experiment.

7. Possible experimental data for the existence of other universes is considered.

8. A model of Hubble expansion of our universe is defined. We call it the HubbaHubble model. It has an interesting Hubble parameter Big Dip that is mathematically required by current experimental data.

9. Fibonacci and Ramsey numbers are considered from a Cosmos Theory perspective.

1. Top Down and Bottom Up Derivations of Cosmos Theory

Our work in 2022-2023 was focused on the derivation and features of Cosmos Theory from a "Top Down" approach.[1] In this book we develop a "Bottom Up" approach to Cosmos Theory that provides a yet deeper view of the basis of Cosmos Theory. We begin with a summary of the Top Down approach and then turn to a Bottom Up approach.

1.1 Top Down Approach to Cosmos Theory

The Top Down approach to Cosmos Theory is described in our books appearing in 2019-2023 with additional material presented in earlier books ranging back to the year 2000. In this section we outline the Top Down approach referring the reader to our earlier books for details.

1. Cosmos Theory begins with two complementary origins:
    a. A derivation of the ten HyperCosmos and ten Second Kind HyperCosmos spectrums of spaces from a Geometric ProtoCosmos Theory[2] using PseudoFermion wave functions.[3]

    AND/OR

    b. A derivation of the ten HyperCosmos and ten Second Kind HyperCosmos spectrums of spaces from an analysis of the creation/annihilation operator structure of the r space-time dimension Fourier expansion of a second quantized fermion wave function. The set of creation/annihilation operators is enlarged by a factor generated by attaching an index representing internal symmetries, whose range is the Cayley[4] number equal to the number of creation/annihilation operators.[5]

The result is a Spectrums of Spaces.

---

[1] The terms "Top Down" and "Bottom Up" originate in Computer Science. They aptly describe the approaches presented here. (They have no explicit Computer Science aspects. Top Down indicates a derivation from a Big Picture perspective down to a detailed description. Bottom Up indicates a "derivation" starting from s detailed primitive description up to a full description.

[2] There are earlier ProtoCosmos models such as the UniDiemsion Models which are noe superseded.

[3] See Blaha (2023c).

[4] Some Cayley numbers are 1, 2, 4, 8, 16, 32, ... **They are powers of 2 and thus correctly enumerate the numbers of wave function creation/annihilation operators, which alwaya number in powers of 2 for fermions in all space-time dimensions.**

[5] Chapter 3 of Blaha (2023a) and chapter 4 of Blaha (2023c) as well as earlier books.

2. Analysis of the features of Cosmos Spaces
   Each of the 20 Cosmos spaces: 10 HyperCosmos spaces and 10 Second Kind HyperCosmos spaces is defined with a set of real-valued space-time coordinates and a set of internal symmetry dimensions, which map to internal symmetry group representations. In themselves the spaces have no form or shape. Each constitutes a template or blueprint for a universe containing mass, energy, particles, symmetry groups, and interactions. Universes are defined with creation/annihilation operators specifying their space and energy. The result:

   a. The total set of coordinates and dimensions of each space form a Dimension Array, which is treated as a square array whose sides number $2C_n$ dimensions where $C_n = 2^n$ is the nth Cayley number.
   b. Fundamental Reference Frames (FRFs) This topic and the following topics in this section are covered in detail in Blaha (2022c), (2022f) and (2023c) as well as our other books.
   c. Transformations of FRFs
   d. Fermion and Symmetry Group Contents of an FRF
   e. Replicates of sets of fermions and internal symmetries for FRFs and Cosmos Theory spaces
   f. Subluminal/Superluminal FRF format of sets of particles and Symmetry Groups

3. Unification Space of each HyperCosmos and Second Kind HyperCosmos Space
   Each of the 20 spaces has a corresponding Unification space whose set of General Relativistic transformations of its fermions wave function creation/annihilation operators unify the space's set of space-time and internal symmetries.

4. Combined Full Unification Space
   The set of 20 HyperCosmos and 2nd Kind HyperCosmos Unification spaces combine to populate the 42 space-time dimension Full Unification space. This space supports combined General Relativistic (GR) transformations of all 20 individual General Relativistic transformations implementing unification within each of the spaces. The 20 individual General Relativistic transformations can be subsumed in a 42 space-time dimension "generalized" General Relativistic transformation that provides a further unification implementing "rotations" of *all* creation/annihilation operators of all 20 spaces. A consequence of the set of generalized GR transformations is the ability to generate all dimensions of the dimension arrays of all 20 spaces from one dimension in the FRF of the Full Unification space.

5. UltraUnification Space

Generation of all dimensions from one primordial dimension

6. Dimension Array Algebra

7. The Unified SuperStandard Theory (UST) of our Universe (and Space)

These features are explored in detail later.

1.2 Bottom Up Approach to Cosmos Theory

This book also develops Cosmos Theory starting from the map between Minkowski space (and higher space-time dimension spaces) and the Spin group su(2, 2). (See section 2.3.) In higher space-time dimensions we find the Spin groups associated with the HyperCosmos and Second Kind HyperCosmos spectrums displayed in Figs. 2.1 and 2.2. The Spin groups of higher space-time dimension spaces have su(2, 2) as a subgroup.

It is important to note that the number of fermion spin states in any space-time dimension is a power of 2. This observation is the basis of the structure of spaces in Cosmos Theory as a fermion-based fundamental theory.

From this basis the bottom up derivation of Cosmos Theory emerges:

1. Spaces should be defined for an energy spectrum expressed in powers of 2.

2. We define a ProtoCosmos space in which there is a Hydrogen-like "atom" whose energy levels form a geometric spectrum.

3. We augment the "atom" by adding a factor to each atom wave function specifying a fermion. The fermion will have a part that is in the ProtoCosmos space (the parent space) and a part that is in a subspace (the child space). Each part may be used to define a universe. We call a fermion wave function of this type an Independent PseudoFermion (IPF) wave function. (Chapter 11) The parent space and the child space each define independent dimension arrays.

4. IPF wave functions enable the child space to generate subspaces due to their independent dimension arrays.

5. Having created a spectrum of spaces where each space has a dimension array, we can proceed to define fermions and symmetry groups for each space.

6. The dimension array of each space may be related to a more compact array through the use of a General Relativistic transformation from a frame that we call a Fundamental Reference Frame (FRF). We view FRFs as the "rest frames" of spaces.

7. The set of dimensions in a FRF can be shown to be the result of a superluminal/subluminal classification of dimensions, fermions, other particles and symmetry groups based on su(1,1). (Chapter 4)

8. We may then define Unification spaces where purely General Relativistic transformations map between a frame and an FRF.

9. Full Unification spaces and an UltraUnification space may then be defined.

This "bottom up" procedure is implemented in the following chapters along with the "top down" procedure which is closely related to it.

2. The Space-Time Basis of Cosmos Theory

This chapter begins the development of a Bottom Up approach to Cosmos Theory. It begins with an analysis of the central role of fermion quantum field theory in structuring the spaces of the Cosmos. It then develops the forms of the fundamental set of fermions and the fundamental set of internal symmetry groups based on a superluminal/subluminal space-time analysis. The following chapters "grow" these fundamental sets to the full grown dimension arrays of Cosmos Theory.

2.1 The Meaning of Space-Time

Space-time is subject to a number of interpretations. The simplest approach is to view space-time dimensions as coordinates that have parameter values. In chapter 13 of Blaha (2022c) we considered this view and specified the rule:

The dimension of a space-time is the number of independent parameters needed to define a point. An over-determined space-time has a negative dimension.

The question arises: How do space-time dimensions differ from the dimensions of representations of internal symmetry groups. Since Cosmos Theory unifies space-times and Internal Symmetries both types of dimensions must be the same. Cosmos Theory General Relativity transformations transform among both types of dimensions to achieve Unification.

Yet a problem still remains. Space-time coordinates are used in the definition of Yang-Mills fields for Internal Symmetries. However internal symmetry representation coordinates are restricted to each internal symmetry group separately. For example, one doesn't normally parameterize a SU(2)⊗U(1) group representation with the SU(3) coordinates in the Standard Model.

It seems the only resolution of this issue resides in the choice of dynamics. We choose a single time coordinate and use it to determine a dynamical evolution in Quantum Theory. Secondly we choose a set of spatial coordinates. In doing this we arbitrarily require only one time coordinate. We also arbitrarily specify the number of spatial coordinates. We require the space and time coordinate values to be real-valued numbers. The set of coordinates can only be viewed as a property of the universe and its associated space.

The space-time coordinates, so chosen, have an associated metric to distinguish between them. The metric of a flat space-time assigns differing metric coefficients such as

$$ds^2 = t^2 - \mathbf{x}^2. \tag{2.1}$$

where t is the time and **x** is the spatial vector. In our universe we have a Minkowski space-time which maps to the Spin group su(2, 2). In other universes of other spaces the associated Spin groups for each space-time are listed in Figs. 2.1 and 2.2.

As a result it appears that the distinguishing factor between space-time and Internal Symmetry dimensions is the dynamical choice of time coordinate and the implied choice of spatial coordinates due to the requirement of Special Relativistic (and General Relativistic) covariance.

2.2 Example of the Separation of Space-time and Internal Symmetry Dimensions

An example that illustrates the separation of space-time and internal symmetries begins with a four quaternion dimension space with 16 independent dimensions (coordinates).[6] The space may be viewed as an su(4, 4) irreducible representation. We break the space to su(2, 2)⊗su(2, 2). Each factor defines a four complex dimension space. We may then proceed to extract an SU(4) symmetry[7] and irreducible representation from one factor. We then extract an SU(2)⊗U(1)⊗SO^+(1,3) symmetry from the other factor. The SO^+(1, 3) factor defines a real-valued four dimension space that we comprises the dynamical space-time. The other factors combine to give the Standard Model symmetries:[8]

$$SU(4)\otimes SU(2)\otimes U(1)\otimes SL(2,\mathbf{C}) \tag{2.2}$$

Thus the four quaternion dimension space[9] separates into four real-valued coordinates' space-time and 12 real-valued internal symmetry dimensions. In the analysis of the spaces of Cosmos Theory we separate each dimension array into a space-time set of dimensions and into sets of dimensions for internal symmetry representations.

2.3 The Central Role of Fermions

Our geometric formulation of the ProtoCosmos and of Cosmos Theory raises the question of the rationale for the role of PseudoFermion wave functions.

In our universe's four dimensions the basis of the critical role of PseudoFermions and fermions is the map from Minkowski space to a twistor-like space based on half integer spin. Its consequence is the use of Dirac γ matrices[10] and spinors with their associated Fourier representation creation/annihilation operators in the formulation of Cosmos Theory. In four space-time dimensions we note

---

[6] The discussion in this subsection summarizes the much more detailed discussion in chapters 26 through 31 of Blaha (2020c) using complex boosts.

[7] Or SU(3)⊗U(1).

[88] SL(2, C) indicates the Lorentz group. SO^+(1 ,3) in particular

[99] This space has both superluminal and subluminal parts. (See Chapter 4.) We have shown that superluminal Physics has no evident problems. Appendix A (from our earlier work) confirms superluminal quantum field theory is acceptable. Appendix B (also from our earlier work) shows that Statistical Mechanics and Thermodynamics are also satisfactory.

[10] We use γ matrices rather than spinors to put negative energy stastes and anti-particles on an equal footing with positive energy states and particles.

$$\text{Minkowski space} \leftrightarrow \text{su}(2, 2) \text{ Spin group} \quad (2.3)$$

In higher space-time dimensions we find the Spin groups in the HyperCosmos and Second Kind HyperCosmos spectrums displayed in Figs. 2.1 and 2.2. The Spin groups of higher space-time dimension spaces have su(2, 2) as a subgroup. *This leads to the result that Cosmos Theory is a fermion-based fundamental theory.*

2.4 The Fundamental Groupings of Creation/Annihilation Operators

Having established the central role of fermions we proceed to examine the creation and annihilation operators in fermion Fourier expansions. We begin by developing an algorithm for the number of degrees of freedom associated with a wave PseudoQuantum[11] wave function in a space-time of r dimensions.[12]

It is important to note that the number of spin states in fermions in any space-time dimension is always a power of 2. This observation is the basis of the structure of spaces in Cosmos Theory.

Each of the two wave functions has 8 operators (namely b_1, $b_1^\dagger$, b_2, $b_2^\dagger$, d_1, $d_1^\dagger$, d_2, $d_2^\dagger$) for each spin. There are $2^{r/2-1}$ spins. Thus taking account of the spin degrees of freedom we find the factor:

$$2^{r/2+2} \quad (2.4)$$

Then we assume that internal symmetries introduce a similar factor[13] yielding a total of

$$d_{dN} = 2^{r+4} \quad (2.5)$$

degrees of freedom, which we take to be the number of dimensions in the dimension array d_{dN} for the space-time of dimension r corresponding to the Nth space in Fig. 2.1.[14]

Eq. 2.5 implies the space-time dimension r is related to the size (number of elements) of the dimension array for HyperCosmos spaces:

$$r = \ln_2 (d_{dN} /16) \quad (2.6)$$

For Second Kind HyperCosmos spaces (Fig. 2.2) we find

$$r = \ln_2 (d_{dN} /8) \quad (2.7)$$

since its dimension arrays satisfy

$$d_{d2N} = 2^{r+3} \quad (2.8)$$

---

[11] There are two PseudoQuantum Fermion fields for a particle introducing a factor of two in eq. 2.4.

[12] See chapter 4 of Blaha (2023c) and earlier books.

[13] The choice of an "internal symmetry" factor equal to the number of spin values is the only simple justifiable choice. Dimension democracy!

[14] This calculation is for HyperCosmos spaces. For Second Kind HyperCosmos spaces the result is 2^{r+3}.

2.5 Space-time Dimensions

The space-time dimensions of the HyperCosmos and Second Kind HyperCosmos spaces are determined by the following considerations:

1. The minimal value of r is zero.
2. The values of rare even integers. The creation/annihilation operators of odd r space-times number the same as those of the r – 1 even dimension. The choice of even r values only eliminates a potential redundancy.
3. The number of spaces is limited to 10 arbitrarily causing the maximum value of r to be 18.

The resulting space spectrums appear in Figs. 2.1 and 2.2.

2.6 HyperCosmos Dimension Array Columns and Sizes

The spin of fermions is related to Cayley numbers. The number of spins of a fermion in r dimensions is

$$2^{r/2-1} \text{ spins} \tag{2.9}$$

The nth number Cayley number C_n in the Cayley-Dickson construction is

$$C_n = 2^n \tag{2.10}$$

Consequently the number of spins is

$$C_{r/2 - 1} \tag{2.10}$$

Since the number of fermion spins is directly related to the size (number of elements) and number of columns of a HyperCosmos dimension array we find:
The corresponding number of elements in a HyperCosmos dimension array column vector[15] is

$$d_{cdn} = 2C_n \tag{2.11}$$

and the HyperCosmos space n^{th} dimension array size is

$$d_{dN} = (2C_n)^2 \tag{2.11}$$

where Blaha number N satisfies

$$N = 10 - n \tag{2.12}$$

We can express d_{dN} with

$$d_{dn} = d_{cdn}^{\ 2} = (2C_n)^2 = 2^{2n+2} \tag{2.13}$$

and

$$d_{dn} = 2^{r+4} \tag{2.14}$$

---

[15] See Fig. 6.1.

The HyperCosmos space spectrum entries in Fig. 2.1 reflect the relation of Cayley numbers to dimension arrays.

The size and number of columns in a Second Kind HyperCosmos dimension array is:

$$d_{cd2n} = 2C_n \tag{2.15}$$

The Second Kind HyperCosmos space n^{th} dimension array (Fig. 2.2) size d_{dN2} is

$$d_{d2N} = 2C_n^{\,2} \tag{2.16}$$

where Blaha number N again satisfies

$$N = 10 - n \tag{2.17}$$

We may express d_{d2N} with

$$d_{d2n} = d_{cd2n}^{\,2}/2 = d_{dn}/2 = 2C_n^{\,2} = 2^{2n+1} \tag{2.18}$$

The Second Kind HyperCosmos space spectrum entries in Fig. 2.2 reflect the relation of Cayley numbers to dimension arrays.

2.7 Subsets of Creation/Annihilation Operators Determining Fermion and Symmetry Group Structure

Within a set of creation and annihilation operators we can identify a set of 8 operators for each spin. We can further identify separate sets of four operators with one set for type 1 PseudoQuantum fields and one set for type 2 PseudoQuantum fields. If we group sets of 8's by positive and negative spin we obtain sets of 16.

The separation into sets of 4, 8 and 16 will have results later in the construction of the contents of the Fundamental Reference Frames.

We note the HyperCosmos space dimension arrays form square arrays of $2^{r/2+2}$ rows and $2^{r/2+2}$ columns. The Second Kind HyperCosmos space dimension arrays form oblong arrays of $2^{r/2+2}$ rows and $2^{r/2+1}$ columns.

The dimension arrays provide sets of dimensions. These dimensions may be structured to form sets of irreducible symmetry group dimensions. They may also be used to determine the structure of the fundamental fermion spectrum since the fundamental fermions play the role of vectors in fundamental group irreducible representations. In addition they also determine scalar and vector boson structures according to the internal symmetry groups of each space.

2.8 Fermion Spectrum Structure

The fermion structure implied by the creation/annihilation operator structuring is:

1. A separation of fermions into sets of four fermions[16] composed of a set consisting of (e, 3 up-quarks) and a set of (ν, 3 down-quarks) of the Normal and Dark types.

---

[16] Assuming an initial SU(4) grouping that is broken to SU(3)⊗U(1) subsequently.

2. The Normal and Dark types consist of 8 fermions.
3. The fermions appear in four generations in four layers in the Unified SuperStandard Theory (UST) shown in Figs. 2.3 and 2.4 for a HyperCosmos space for our universe.

See Figs. 2.3 and 2.4. The structuring implied by the structure of the creation/annihilation operators directly appears in the fermion spectrum for our N = 7 space's universe.

Since the higher space-time dimension spaces are composites in sets of four spaces of the next lower (by two dimensions) space-time dimension the pattern of Figs. 2.3 and 2.4 are nested in these higher dimension fermion spectrums. Thus the structuring applies throughout the 10 spaces and their universes.

The fermion structure of Second Kind HyperCosmos spaces is the same with the only difference being the absence of Dark fermions. Our universe may be a HyperCosmos universe or a Second Kind HyperCosmos universe. The key factor is the presence or absence of a Dark sector – an experimental question!

2.9 Symmetry Groups Structure

Symmetry group structure is also implied by the creation/annihilation operator structure. The symmetry group structuring for our N = 7 space's universe is:

1. An initial separation of the dimensions in the dimension array into sets of eight real-valued U(4) irreducible representations as shown in Fig. 2.5 for the N = 7 space of our universe (UST).

2. Note a pair of two U(4) groups – 16 dimensions appearing twice in each layer in Fig. 2.5.

3. The U(4) groups are broken to familiar groups of the Standard Model and UST.

4. One U(4), that of the lowest layer, becomes SU(2)⊗U(1)⊗SL(2, **C**) where the SL(2, **C**) is for the space-time coordinates. Seven other U(4) groups become SU(2)⊗U(1)⊗SU(2) groups that implement Connection groups connecting Normal and Dark layers as shown in earlier books. See Fig. 2.7 for the Connection groups for our universe.

5. Each of the four layers has its own symmetry groups.

See Figs. 2.5, 2.6 and 2.7. The structuring implied by the structure of the creation/annihilation operators directly appears in the set of symmetry groups fermion spectrum for our N = 7 space's universe.

Since the higher space-time dimension spaces are composites formed as sets of four spaces of the next lower space-time dimension's dimension array the patterns of Figs. 2.5 and 2.6 are nested in these higher dimension Internal Symmetry groupings. Thus the internal symmetry structuring applies throughout the 10 spaces and their universes.

The symmetry group structure of Second Kind HyperCosmos spaces is the same with the only difference being the absence of Dark sector. Our universe may be a HyperCosmos universe or a Second Kind HyperCosmos universe. The key factor again is the presence or absence of a Dark sector.

2.10 Scalar and Vector Bosons Structure

The scalar (including Higgs) and vector bosons of each space exactly parallels that of symmetry groups since they lie in irreducible representations of the symmetry groups as presented in section 2.9.

2.11 Higher Space-time Dimension Spaces

Higher space-time dimension spaces dimension arrays with N < 7 are constructed from the dimension arrays of lower spaces due to Cayley numbers compounding as in the Cayley-Dickson construction. For example the N = 6 (Multiverse/Megaverse) space dimension array is a four-fold duplication of the N = 7 dimension array. For the HyperCosmos N = 6 space the dimension array is a square array consisting of four N = 7 dimension arrays. Consequently the above comments for the N = 7 case carry over to the N = 6 case duplicate parts.

For the Second Kind HyperCosmos space case the N = 6 dimension array is an oblong array consisting of four oblong N = 7 dimension arrays.[17] Consequently the above comments for the N = 7 Second Kind HyperCosmos case carry over to the N = 6 case duplicate parts.

The same type of fourfold duplication take place space by space as one goes up the spectrum. It applies to the entire spectrums of these HyperCosmos and Second Kind HyperCosmos spaces.

2.12 Bottom Up HyperCosmos and Second Kind HyperCosmos Space Spectrums

The above Bottom Up derivation, supplemented with details from our earlier books, provides a complete presentation of these Cosmos Theory spaces. The derivation starts with the form of its creation/annihilation operators and leads directly to the fermion and symmetry group structure of each of the twenty spaces.

---

[17] These oblong arrays have the horizontal row sizes one-half the sizes of the vertical column sizes. The result is the absence of Dark sectors of fermions and symmetries.

THE HYPERCOSMOS SPACES SPECTRUM

Blaha Space Number $N = o_s$	Cayley-Dickson Number n	Cayley Number Cn d_c	Dimension Array column length d_{cd}	Dimension Array Size d_{dN}	Space-time-Dimension r	CASe Group $su(2^{r/2}, 2^{r/2})$ CASe
0	10	1024	2048	2048^2	18	su(512,512)
1	9	512	1024	1024^2	16	su(256,256)
2	8	256	512	512^2	14	su(128,128)
3	7	128	256	256^2	12	su(64,64)
4	6	64	128	128^2	10	su(32,32)
5	5	32	64	64^2	8	su(16,16)
6	4	16	32	32^2	6	su(8,8)
7	**3**	**8**	**16**	$\mathbf{16^2}$	**4**	**su(4,4)**
8	2	4	8	8^2	2	su(2,2)
9	1	2	4	4^2	0	su(1,1)
10	0	1	2	2^2	-2	
11	-1	½	1	1	-4	

Figure 2.1. The HyperCosmos space spectrum augmented with N = 10 and N = 11 lines for use later. (Spaces with negative space-times may have universes. See Blaha (2022c).)

HYPERCOSMOS OF THE SECOND KIND SPACES SPECTRUM

Blaha Space Number $N = O_s$	Cayley-Dickson Number n	Cayley Number d_c	Dimension Array size d_{dN2}	Space-time-Dimension r	CASe Group $su(2^{r/2}, 2^{r/2})$ CASe
0	10	1024	1024 × 2048	18	su(512,512)
1	9	512	512 × 1024	16	su(256,256)
2	8	256	256 × 512	14	su(128,128)
3	7	128	128 × 256	12	su(64,64)
4	6	64	64 × 128	10	su(32,32)
5	5	32	32 × 64	8	su(16,16)
6	4	16	16 × 32	6	su(8,8)
7	**3**	**8**	**8 ×16**	**4**	**su(4,4)**
8	**2**	4	4 × 8	2	su(2,2)
9	**1**	2	2 × 4	0	su(1,1)
10	0	1	1 × 2	-2	
11	-2	½	½	-4	

Figure 2.2. The HyperCosmos of the Second Kind space spectrum augmented with N = 10 and N = 11 lines for use later. (Spaces with negative space-times may have universes.)

Number of Columns =	4 NORMAL 4		4 DARK 4	
Layer 1 4 generations	e 3 up-quarks	ν 3 down-quarks	e 3 up-quarks	ν 3 down-quarks
Layer 2 4 generations	e 3 up-quarks	ν 3 down-quarks	e 3 up-quarks	ν 3 down-quarks
Layer 3 4 generations	e 3 up-quarks	ν 3 down-quarks	e 3 up-quarks	ν 3 down-quarks
Layer 4 4 generations	e 3 up-quarks	ν 3 down-quarks	e 3 up-quarks	ν 3 down-quarks

Figure 2.3. The N = 7 QUeST 16 × 16 fermion spectrum of our universe tentatively arranged as SU(4)-plets that correspond directly with SU(4) (or SU(3)⊗U(1)) fermions. There are four layers. Each set of 4 fermions has 4 generations matching the number of rows in each layer. This Periodic Table is broken into Normal and Dark sectors.

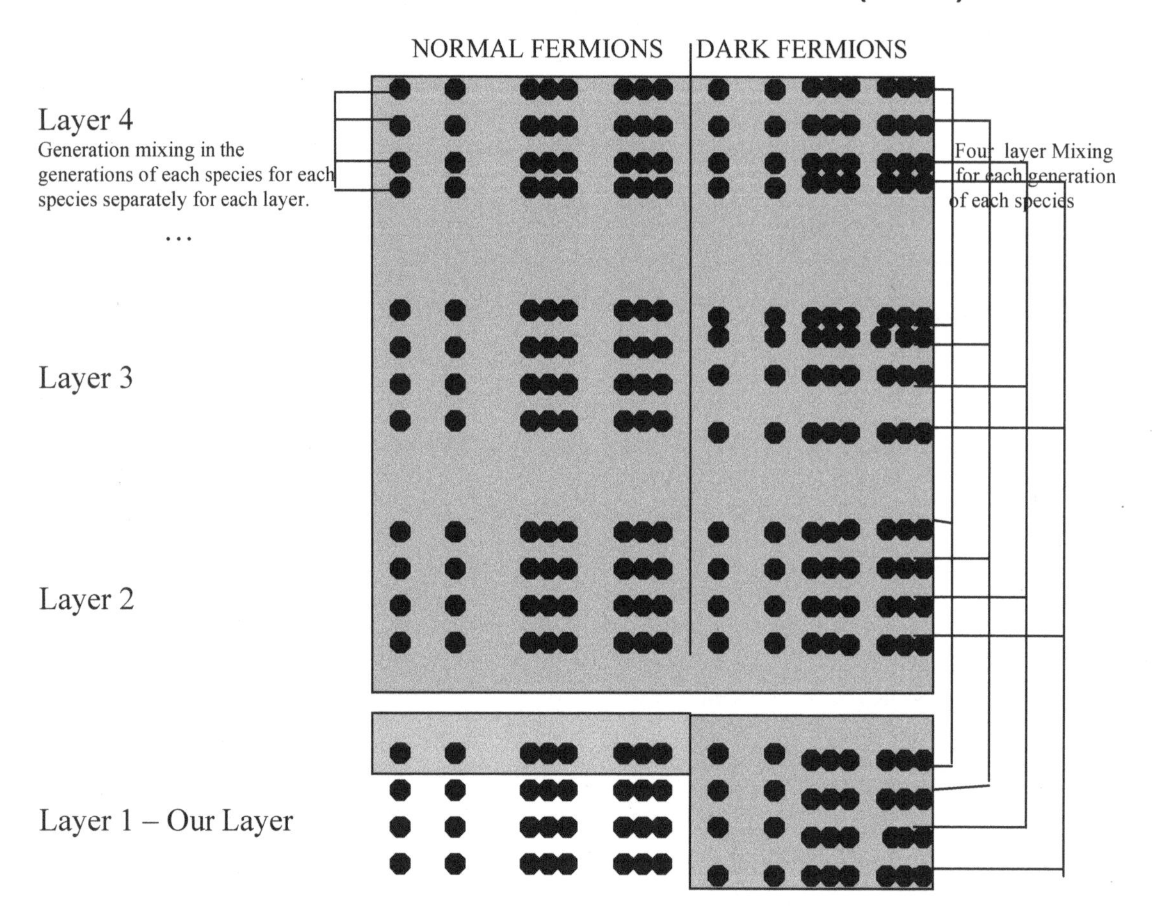

Figure 2.4. The N = 7 QUeST Fermion particle spectrum and partial examples of the pattern of mass mixing of the Generation groups and of the Layer groups. Unshaded parts are the known fermions, A shaded generation is an additional, as yet not found, 4th generation. The lines on the left side (only shown for one layer) display the Generation Group mixing within each layer. The Generation mixing occurs within each layer using a separate Generation group for each layer. The lines on the right side show Layer group mixing (for Dark matter) with the mixing among all four layers for each of the four generations individually. See Chapter 10 for details. There are four Layer groups for Normal matter and four Layer groups for Dark matter. There are 256 fundamental fermions. QUeST and UST have the same fermion spectrum.

	NORMAL	DARK
Layer 1	U(4)⊗U(4) U(4)⊗U(4)	U(4)⊗U(4) U(4)⊗U(4)
Layer 2	U(4)⊗U(4) U(4)⊗U(4)	U(4)⊗U(4) U(4)⊗U(4)
Layer 3	U(4)⊗U(4) U(4)⊗U(4)	U(4)⊗U(4) U(4)⊗U(4)
Layer 4	U(4)⊗U(4) U(4)⊗U(4)	U(4)⊗U(4) U(4)⊗U(4)

Figure 2.5. The "initial" distribution of sets of N = 7 symmetry groups. Each of the four sets is distinct and supports interactions only for the corresponding set of fermions in its layer (separately for Normal and Dark fermions). *Thus each set of 4 fermion generations has its own quantum numbers and interactions.* Each U(4)⊗U(4) set has a 16 real-valued dimension representation. The Layer group representations are spread over all four layers. See Chapter 10 for details.

NORMAL		DARK	
SU(3)⊗U(1) Generation U(4)	SU(2)⊗U(1)⊗SL(2, C) Layer U(4)	SU(3)⊗U(1) Generation U(4)	SU(2)⊗U(1)⊗U(2) Layer U(4)
SU(3)⊗U(1) Generation U(4)	SU(2)⊗U(1)⊗U(2) Layer U(4)	SU(3)⊗U(1) Generation U(4)	SU(2)⊗U(1)⊗U(2) Layer U(4)
SU(3)⊗U(1) Generation U(4)	SU(2)⊗U(1)⊗U(2) Layer U(4)	SU(3)⊗U(1) Generation U(4)	SU(2)⊗U(1)⊗U(2) Layer U(4)
SU(3)⊗U(1) Generation U(4)	SU(2)⊗U(1)⊗U(2) Layer U(4)	SU(3)⊗U(1) Generation U(4)	SU(2)⊗U(1)⊗U(2) Layer U(4)

Figure 2.6. The transformed/broken sets of symmetries in QUeST (UST) and in N = 7 HyperCosmos space.. Note each element has a 16 real dimension representation. This depiction is also evident in QUeST and the UST. The SL(2, **C**) representation has four coordinates.[18]

[18] The Lorentz Group $SO^+(1, 3)$ is often specified with an SL(2, **C**) representation.

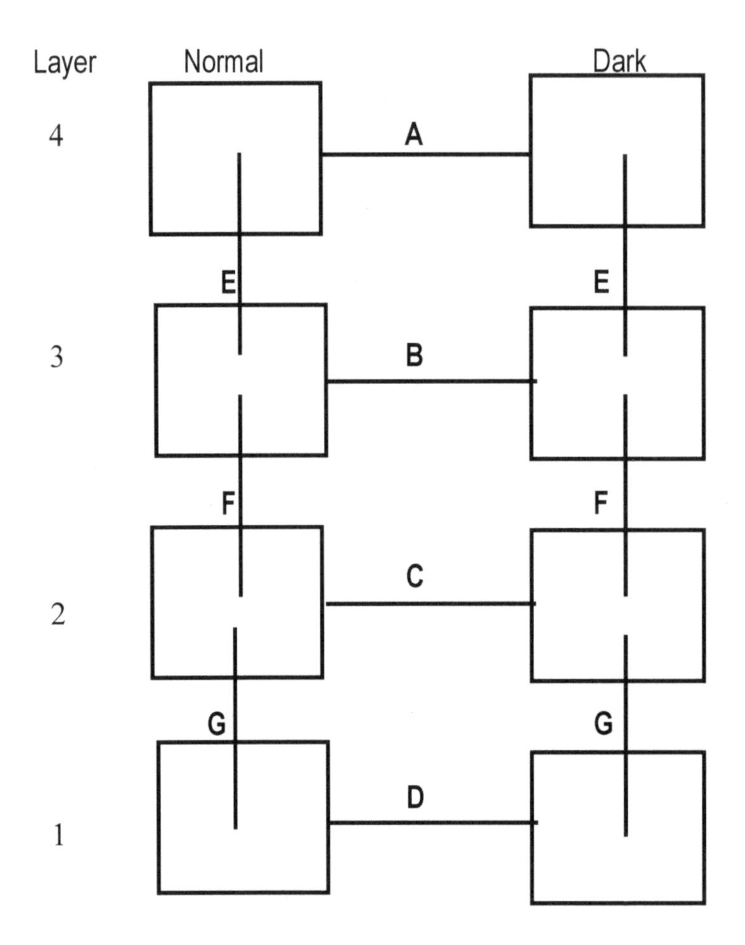

Figure 2.7. The seven U(2) Connection groups[19] between the eight QUeST/UST blocks in the N = 7 HyperCosmos. Connection groups are obtained by transfering dimensions from initial QUeST space-time to internal symmetries. The E, F, G groups are the same for the Normal and Dark sectors. See Appendix 9-B for details.

[19] Connection groups are discussed in Appendix 9-B.

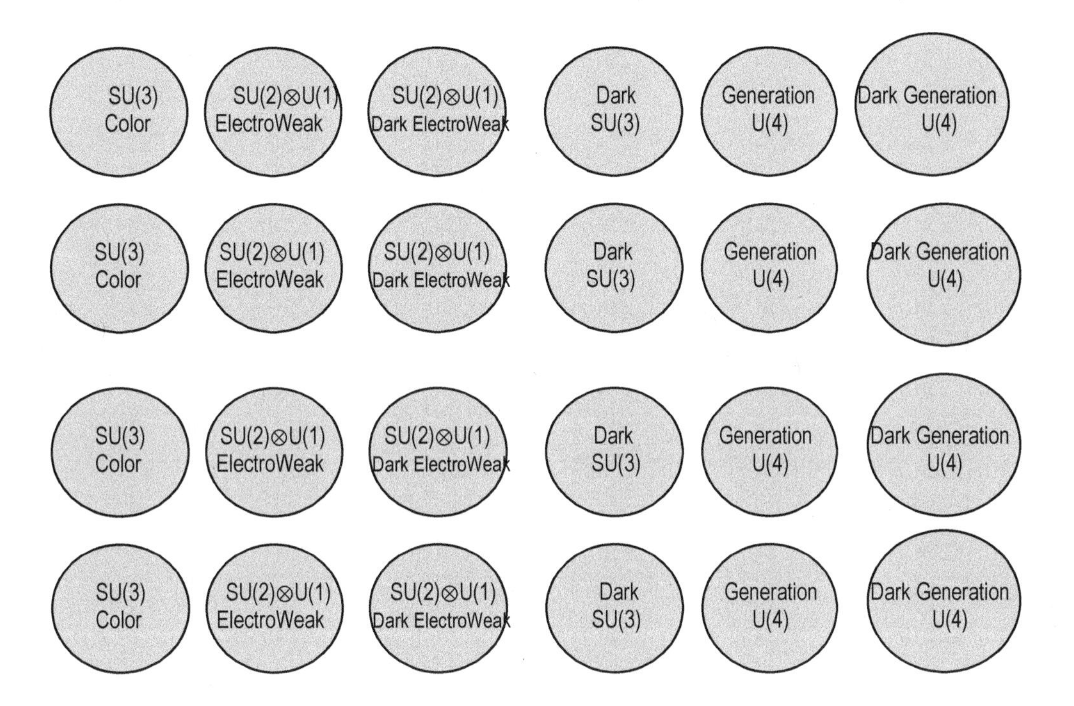

Figure 2.8. (Fig. 2.2 of Blaha (2020a)) The QUeST set of four layers of internal symmetry groups corresponding to four generations in four layers of spin ½ fermions. In addition there are the Normal and Dark Layer groups,

3. The Fundamental Reference Frame (FRF) and its Contents

In chapters 6 and 7 of Blaha (2021j) we showed how to define composite General Relativistic (GR) – Internal Symmetry transformations in a space from a subset of a dimension array elements to the full set of the dimension array using a sedenion formalism in the case of the N = 7 space of our universe. We introduced the concept of a Fundamental Reference Frame (FRF). In this chapter we describe the process of transforming a space's dimensions (and fermions and Internal Symmetries) from an FRF to a dimension array in the space where the transformation is a strictly GR transformation in the space's HyperUnification space. The set of these GR transformations implement unification.

The Fundamental Reference Frames of the various spaces are determined from their space-time dimension r. It begins with the definition of a HyperUnification space. A HyperUnification space is a space that supports General Relativistic (GR) transformations of its dimension array that transform both space-time and internal symmetry dimensions. Thus it provides a unification of space-time and internal symmetries. This transformation of dimensions is directly analogous to the transformation of creation/annihilation operators of fermion wave functions.[20] The elements of the dimension array of a space become the components in the column vector of the hyperUnification space that undergoes GR transformations. Thus we see the progression:

Space dimension Array →
HyperUnification dimension array column vector →
GR Transformation →
New HyperUnification dimension array column →
New Space Dimension Array

As a result the space's dimension array is transformed mixing both space-time and internal symmetry dimensions.

A HyperUnification space for a HyperCosmos or Second Kind HyperCosmos space has a dimension array whose column length equals the total number of dimensions in the space's dimension array. The HyperUnification space's dimension array satisfies either

$$d_{dN} = 2^{r' + 4} \qquad \text{HyperCosmos space} \tag{3.1}$$

or

$$d_{d2N} = 2^{r' + 3} \qquad 2^{nd} \text{ Kind HyperCosmos space} \tag{3.2}$$

---

[20] First presented in Stephen Blaha, "The Local Definition of Asymptotic Particle States", IL Nuovo Cimento **49A**, 35 (1979) by the author. PseudoQuantum field Theory appears in Blaha (2021j) as well as in S. Blaha, Phys. Rev. **D17**, 994 (1978).; S. Blaha, "New Framework for Gauge Field Theories", IL Nuovo Cimento **49A**, 113 (1979).

The corresponding HyperUnification of space-time dimension r' space's dimension array has its dimension array column length satisfying

$$d_c = 2^{r'/2 + 2} \qquad \text{HyperUnification HyperCosmos space} \tag{3.3}$$

implying

$$r'/2 + 2 = r + 4$$

or

$$r' = 2r + 4 \tag{3.4}$$

For a 2nd Kind HyperCosmos space we find

$$d_{2c} = 2^{r'/2 + 1} \qquad \text{2nd Kind HyperUnification HyperCosmos space}^{21} \tag{3.5}$$

implying

$$r'/2 + 1 = r + 3$$

or

$$r' = 2r + 4 \tag{3.6}$$

The role of a HyperUnification space is to support General Relativistic transformations of its dimension array that transform all elements of a space's dimension array[22] *– thus achieving a unification of space-time and internal symmetry.*

The space-time dimension values of all HyperUnification spaces are listed in Figs. 3.1 and 3.2 in the r' column.

3.1 Fundamental Reference Frame (FRF) Physical Interpretation

A common feature of calculations in Physics is to choose a reference frame which facilitates computation. Common choices are the rest frame and the center of mass frame. In this chapter we find a Fundamental Reference Frame (FRF) (actually an infinite set of such frames) that plays a similar role for GR coordinate reference frames.

The set of transformations are strictly HyperUnification space GR transformations that map from an FRF to a "normal static" reference frame. We are motivated in this endeavor by the observation made in previous books (and later) that the fermion particles of a space (universe) and the symmetry groups of a space appear to be each expressible as a set of replicated fermions and symmetry groups respectively. The replications are generated by transformations of an FRF set to a set of replicates.

The analogy to Special Relativistic transformations from a rest frame to a moving frame is clear.

---

[21] The 2nd Kind HyperCosmos and HyperUnification spaces are oblong with r'/2 + 1 colums and r'/2 + 2 rows.

[22] The space's dimension array elements become the elements of the HyperUnification space's dimension array column.

3.2 FRF Contents

We describe the possible contents of an FRF in this section.

3.2.1 FRF Dimensions

An FRF can be defined to have a number of non-zero dimensions that, after GR transformation, become the dimensions of a dimension array. There are two interesting possibilities. Note the r' space-time dimension HyperUnification space dimension array has a column length of $2^{r'/2 + 2}$ and a Second Kind HyperUnification space has a column length of $2^{r'/2 + 1}$.

1. One non-zero dimension

   In this case a GR transformation may generate a column of $2^{r'/2 + 2}$ (or $2^{r'/2 + 1}$.in the case of 2nd Kind HyperCosmos spaces) non-zero elements with the rest equal to zero, which are the elements of the r space-time dimension HyperCosmos dimension array of $2^{r + 4}$ elements by eq. 3.4. For the Second Kind case a dimension array of $2^{r + 3}$ elements may be generated by eq. 3.6.

   Thus one nonzero FRF dimension may be transformed to a completely non-zero dimension array. Note the numeric value of the dimension array elements is irrelevant. Only its zero or nonzero status is relevant.

2. $2^{r/2 + 2}$ non-zero dimensions

   In this case a GR transformation may generate a column of $2^{r'/2 + 2}$ (or 2nd Kind case: $2^{r'/2 + 1}$) non-zero elements with the rest equal to zero, which are the elements of the r space-time dimension HyperCosmos dimension array of 2^{r+4} (2^{r+3}) elements by eq. 3.4. Thus the HyperUnification space GR transformation "multiplies" the number of elements by a factor of $2^{r/2 + 2}$ to create the space dimension array elements.

   For the Second Kind case a dimension array of $2^{r + 3}$ elements may be generated by eq. 3.6 from $2^{r/2 + 1}$ nonzero FRF elements.

   Thus $2^{r/2 + 2}$ nonzero FRF dimensions (the number of entries in the space's square dimension array column vector) may be transformed to a completely non-zero space dimension array in the HyperCosmos case.

   And $2^{r/2 + 1}$ nonzero FRF dimensions (the number of entries in the space's oblong dimension array column vector) may be transformed to a completely non-zero space dimension array in the Second Kind HyperCosmos case.

3. Other FRF Dimension Content

   The FRF content may be defined to have some other set of non-zero dimension content. Again a GR transformation produces the space's dimension array.

Note the numeric value of the dimension array elements is irrelevant. Only its zero or nonzero status is relevant.

3.2.2 FRF Contents

Case 2 above has an FRF content that may be decomposed into subunits that will become important later.

We consider the HyperCosmos space case first. There are $2^{r/2\ +\ 2}$ non-zero dimensions in the HyperUnification vector. For r = 0 there are $d_{cr=0} = 4$ non-zero dimensions. As r increases by increments of 2 we see

$$d_{cr} = 2^{r/2+2} \tag{3.7}$$

For r ≥ 4 the number of *units of eight nonzero dimensions* for HyperCosmos HyperUnification space is

$$d_{cr8} = 2^{r/2-1} \tag{3.8}$$

If r = 4 (our space and universe) d_{cr} = 16 or two units of eight nonzero dimensions. For a Multiverse/Megaverse r = 6 and its space has d_{cr} = 32 or four units of eight nonzero dimensions.[23]

Next, the Second Kind HyperCosmos space case gives: $2^{r/2\ +\ 1}$ non-zero dimensions in the HyperUnification vector. For r = 0 there are $d_{2cr=0} = 2$ non-zero dimensions. As r increases by increments of 2 we see

$$d_{2cr} = 2^{r/2+1} \tag{3.9}$$

For r ≥ 4 we the number of *units of eight nonzero dimensions* is

$$d_{2cr8} = 2^{r/2-2} \tag{3.10}$$

If r = 4 (a Second Kind space and universe) d_{2cr} = 8 or one unit of eight nonzero dimensions. For a 2nd Kind HyperCosmos Multiverse/Megaverse r = 6 and its space has d_{2cr} = 16 or two units of eight nonzero dimensions.

3.2.3 GR Transformation from FRF to Space Dimension Array

For a HyperCosmos space, a GR transformation of an FRF vector in a space's HyperUnification space to the contents of the space's dimension array takes its $2^{r'/2\ +\ 2}$ dimensions (including $2^{r/2\ +\ 2}$ nonzero dimensions) of the vector and replicates them to form a vector consisting of the $2^{r\ +\ 4}$ dimensions of the space's dimension array. The process, which is described at the beginning of this chapter, is depicted as

Space dimension Array →
HyperUnification dimension array column vector →
GR Transformation →
New HyperUnification dimension array column →
New Space Dimension Array

The transformation causes each of the $2^{r/2\ +\ 2}$ nonzero dimensions to be replicated by a factor of $2^{r/2+2}$ to produce the 2^{r+4} dimensions of the space's dimension array.

Consequently the space's dimension array has 2^{r+1} units of 8 dimensions.

---

[23] The unit numbering is directly extendable to r < 4. Eqs. 3.7 – 3.10 hold for r = 2 and r = 0.

For a Second Kind HyperCosmos space, a GR transformation of an FRF vector in a space's HyperUnification space to the contents of the space's dimension array takes its $2^{r/2+2}$ dimensions (including $2^{r/2+1}$ nonzero dimensions) of the vector and replicates them to form a vector consisting of the 2^{r+3} dimensions of the space's dimension array. The process, which is described at the beginning of this chapter, is depicted as

Space dimension Array →
HyperUnification dimension array column vector →
GR Transformation →
New HyperUnification dimension array column →
New Space Dimension Array

The transformation causes each of the $2^{r/2+1}$ nonzero dimensions to be replicated by a multiple of $2^{r/2+2}$ to produce the 2^{r+3} dimensions of the space's dimension array.

Consequently the Second Kind HyperCosmos space's dimension array has 2^r units of 8 dimensions. The Second Kind HyperCosmos space does not have a Dark sector.[24]

3.3 FRF Fermion Content

Turning now to the fundamental fermions of our N = 7 space's universe we note that for our HyperCosmos space's universe there are two units of eight dimensions. If we map each dimension to a fundamental fermion and associate one unit of eight dimensions with an SU(4) quadruplet of Normal fermions e q_1 q_2 q_3 and associate one unit of eight dimensions with an SU(4) quadruplet of Dark fermions v dq_1 dq_2 dq_3 then we have a HyperUnification vector of fermions that transforms to the fermion spectrum of Fig. 2.4 containing 16 replicates of the pairs of 8 dimensions giving the Normal and Dark fermion sectors totaling to 256 fermions.

For the fundamental fermions of an N = 6 space's Multiverse/Megaverse we note that there are four units of eight dimensions. If we map each dimension to a fundamental fermion and associate two units of eight dimensions with two SU(4) quadruplets of Normal fermions e q_1 q_2 q_3 and associate two units of eight dimensions with two SU(4) quadruplets of Dark fermions v dq_1 dq_2 dq_3 then we have a HyperUnification vector of fermions that transforms to the fermion spectrum of Fig. 14.3 of Blaha (2020a). There are 32 replicates of the four units of fermions totaling to 1024 fermions.

These fermion spectrums (and internal symmetry groups) for the hypercomplex theories QUeST and MOST led the author, ***upon seeing the close similarity to his Unified SuperStandard Theory (UST),*** *to his 3.5 year effort to develop Cosmos Theory thus leading to this book. The map from dimensions to fermions was provisionally explained by a Quantum Functionals extension of Quantum Field Theory also described in Blaha (2020a).*

For spaces with r > 4 the fermion spectrum may be viewed as composed as quadruples as one increases r by twos: 4, 6, 8, … The r = 6 space has four fermion

---

[24] Defining the Second Kind HyperCosmos space without a Dark sector appears to be the simplest way to eliminate half of the dimensions that appear in the corresponding HyperCosmos space dimension array.

spectrum replicates of the r = 4 spectrum, etc. All fermions are distinct and have different quantum numbers and lie in different internal symmetry group representations.

The fermion content of FRF's is derived from superluminal/sublight su(1, 1) group considerations in chapter 4.

3.4 HyperCosmos Space FRF Internal Symmetry Group Content for N = 7 Space

The Internal Symmetry group structure parallels that of the fundamental fermions. The fundamental fermions that we know of experimentally – The Standard Model fermions, lie in irreducible representations of the Internal symmetry groups. Therefore we anticipate broken[25] SU(4) or SU(3)⊗U(1), SU(2)⊗U(1) as suggested in section 2.2.

For the N = 7 space of our universe we now specify the 16 non-zero dimension set in the FRF as two U(4) internal symmetry groups. These groups are broken to an eight dimension SU(4) group and an eight dimension SU(2)⊗U(1)⊗SL(2, **C**) group.

In chapter 4 we provide a rationale for the choice of these groups based on the discussion in section 2.2 and on a consideration of an su(1, 1) space's two times. The group su(1, 1) appears naturally in the creation/annihilation operators discussion presented earlier.

The 16 dimension set of FRF groups is replicated eight times into four U(4)⊗U(4) groups and four times into SU(2)⊗U(1)⊗SL(2, **C**) groups by a GR transformation to a set of groups with a total of 256 irreducible representation dimensions. The set of groups is listed in Figs. 2.5, 2.6 and 2.7. The set is further transformed to have only one set of space-time coordinates and 7 SU(2) groups, which we call Connection groups,[26] that support interactions between layers and Normal and Dark sectors as shown in Fig. 2.7 (the QUeST/UST internal symmetry groups from Fig. 2.2 of Blaha (2020a)).

All replicated groups are independent, having their own representations and gauge vector bosons.

3.5 HyperCosmos Space FRF Internal Symmetry Content for r ≥ 4 Spaces

The irreducible representation dimensions of the groups in the r space-time dimension case FRF occupy $2^{r/2 - 1}$ units of 8 dimensions (Eq. 3.8). Initially we view them as a product of $2^{r/2 - 1}$ U(4) irreducible representations. Thereafter we see breaking of some U(4)⊗U(4) groups to SU(2)⊗U(1)⊗SL(2, **C**) groups.

The $2^{r/2 + 2}$ dimension set of FRF groups are replicated by a GR transformation into a set of groups with $2^{r + 4}$ total irreducible representation dimensions. The set of groups is a multiple of the groups listed in Figs. 2.5 and 2.6 plus additional Connection groups. See Appendix 9-B for a discussion of Connection groups.

All replicated groups are independent, having their own representations and gauge vector bosons.

---

[25] We assume that the Cosmos is free of interactions while constructing Cosmos Theory. We bring in interactions later. See chapter 10.

[26] Appendix 9-B.

3.6 Second Kind HyperCosmos Space FRF Internal Symmetry Content for r ≥ 4 Spaces

The Second Kind HyperCosmos FRF cases parallel the HyperCosmos cases above. For each space the fermion spectrum and the symmetry groups are the same as that of the HyperCosmos except that the Dark sector is absent.

The Symmetry Group content of FRF's is derived from superluminal/sublight su(1, 1) group considerations in chapter 4.

THE HYPERCOSMOS SPACES SPECTRUM

Blaha Space Number $N = O_s$	Cayley-Dickson Number n	Cayley Number d_c	Dimension Array column length d_{cd}	Dimension Array Size d_{dN}	Space-time-Dimension r	Higher Space-time r' Source of d_{dN} r'	Higher Space Array d_{dN}'
0	10	1024	2048	2048^2	18	40	2^{44}
1	9	512	1024	1024^2	16	36	2^{40}
2	8	256	512	512^2	14	32	2^{36}
3	7	128	256	256^2	12	28	2^{32}
4	6	64	128	128^2	10	24	2^{28}
5	5	32	64	64^2	8	20	2^{24}
6	4	16	32	32^2	6	16	2^{20}
7	**3**	**8**	**16**	$\mathbf{16^2}$	**4**	**12**	$\mathbf{2^{16}}$
8	2	4	8	8^2	2	8	2^{12}
9	1	2	4	4^2	0	4	2^{8}

Figure 3.1. The HyperCosmos spaces spectrum related to a unification higher space space-time dimension and its dimension array. Note the spaces with $r' > 18$ are outside the HyperCosmos set of 10 spaces. However they have the same form as the 10 HyperCosmos spaces.

SECOND KIND HYPERCOSMOS SPACES SPECTRUM

Blaha Space Number $N = O_s$	Cayley-Dickson Number n	Cayley Number d_c	Dimension Array column length d_{cd}	Dimension Array Size d_{dN}	Space-time-Dimension r	Higher Space-time r' Source of d_{dN} r'	Higher Space Array d_{dN}'
0	10	1024	2048	$2048^2/2$	18	40	2^{43}
1	9	512	1024	$1024^{22}/2$	16	36	2^{39}
2	8	256	512	$512^{22}/2$	14	32	2^{35}
3	7	128	256	$256^{22}/2$	12	28	2^{31}
4	6	64	128	$128^{22}/2$	10	24	2^{27}
5	5	32	64	$64^{22}/2$	8	20	2^{23}
6	4	16	32	$32^{22}/2$	6	16	2^{19}
7	**3**	**8**	**16**	$\mathbf{16^{22}/2}$	**4**	**12**	$\mathbf{2^{15}}$
8	2	4	8	$8^{22}/2$	2	8	2^{11}
9	1	2	4	$4^{22}/2$	0	4	2^{7}

Figure 3.2. The Second Kind HyperCosmos spaces spectrum.

4. FRF Contents Determined by Two Time Dimensions in HyperUnification Space

The existence of Superluminal dynamics is an existing issue that has been theoretically studied over a period of years. We studied superluminal Quantum Field Theory a number of times. Blaha (2007a) describes superluminal Quantum Field in some detail and shows it defines a completely reasonable dynamics. Blaha (2018e) studies superluminal Statistical Mechanics and Thermodynamics and shows that they conform to Physical expectations such as Maxwell-Boltzmann theory and the laws of Thermodynamics. Appendices A, B and C describe superluminal Physics.

In this chapter we consider the role of superluminal/sublight considerations in the determination of the structure of fermions and Symmetry Groups of Fundamental Reference Frames and in the form of Higgs and other scalar particles.

4.1 The Higgs Mechanism – Quark Confinement Dichotomy

A possibly deeper view of the Higgs Mechanism is based on a Dichotomy between the Higgs Mechanism and Quark confinement. *The Higgs Mechanism and the confining Strong Interaction are alternatives.* The Higgs Mechanism gives masses to particles. Quark Confinement is based on an effective infinite potential (mass) energy that prevents quark deconfinement.

The Higgs Mechanism uses a tachyonic scalar particle implementation[27] for symmetry breaking because it works despite its being superluminal due to negative mass squared. The Strong Interaction has a non-Higgs, non-tachyonic confinement mechanism.[28] These mechanisms appear interrelated.

A simple boson example illustrating the dichotomy in the mechanisms starts with the PseudoQuantum scalar field equations:

$$\Box\varphi_1 - m^2\varphi_1 + g\varphi_1^{\,4} = 0 \tag{4.1}$$
$$\Box\varphi_2 - m^2\varphi_2 + g\varphi_2^{\,4} = 0 \tag{4.2}$$

with tachyonic negative mass squared. Both the type 1 and type 2 fields have the same field equation as we have seen in earlier work.

On the other hand, the prototype Strong Interaction PseudoQuantum scalar field equations:

$$\Box\varphi_1 \pm m^2\varphi_2 = 0 \tag{4.3}$$
$$\Box\varphi_2 = J \tag{4.4}$$

with the quartic interaction dropped for simplicity yield

---

[27] The tachyonic nature is removed by quartic interactions.

[28] See Blaha (2022f) for a detailed discussion of quark confinement and of related MOND-like gravitation.

$$\Box^2\varphi_1 \pm m^2J = 0 \qquad (4.5)$$

The differential operator $\Box^2$ is quartic. Choosing the minus sign to have a positive J source term: results in

$$\Box^2\varphi_1 = m^2J \qquad (4.6)$$

which specifies a confining linear r potential when used in a gauge field theory.[29]

Note the minus sign places eqs. 4.1 and 4.3 on the same footing with the difference that eq. 4.1 is tachyonic but eq.4.3 is not tachyonic due to the appearance of the type 2 field in the eq. 4.3 mass term. Tachyon behavior is avoided in order to obtain Strong interaction confinement.

We called the difference between the Strong Interaction mechanism and the Higgs Mechanism the *Higgs – Confinement Dichotomy* in our earlier books.

4.2 Sublight vs. Superluminal Sectors in su(1,1) and F-Theory

F-Theory and su(1,1), which appear in the HyperCosmos and the Second Kind HyperCosmos, have two time dimensions that lead to a more intricate form of sublight-superluminal separation. These separations lead to the split of symmetries seen in The Standard Model, the UST, and in HyperCosmos spaces. The su(1, 1) group appears as a subspace in all spaces $N \leq 7$ above our $N = 7$ space.[30]

This phenomena together with the structuring imposed by Cosmos Theory almost completely specifies the separations of Internal Symmetry in the Standard Model and its extension in the UST and the HyperCosmos spaces.

The split is evident in the contents of the symmetry group FRF content of each Cosmos Theory space. See sections 3.4 – 3.7. Together with the spaces, HyperUnification spaces and FRFs development in chapter 3 the total structure of the Cosmos Theory universes (including our own UST universe) is determined.

We now proceed to determine the separation of FRF content from superluminal/subluminal considerations.

4.3 Two Times Coordinates and Fermion Separations

An su(1, 1) group coordinate space has a metric with two real time coordinates:

$$ds^2 = t_{01}{}^2 + t_{02}{}^2 - x_1{}^2 - x_2{}^2 \qquad (4.7)$$

In Blaha (2007b) we showed that the four species of fermions: e-type, ν-type, q-up-type and q-down-type followed from the four types of boosts of the complexified Lorentz Group $SO^+(1,3)$, which is often expressed as a representation of SL(2, **C**)). Complex Lorentz group boosts separate into sublight, superluminal, complex sublight, and complex superluminal boosts yielding the four fermion species respectively.

---

[29] See Stephen Blaha, "New Framework for Gauge Field Theories", IL Nuovo Cimento **49A**, 113 (1979) for an SU(3) gauge theory version with confinement.

[30] See section 2.2.

Now we have a more interesting situation with two time coordinates. Each has its own light speed. We take them to be numerically equal, although different "coordinate system-wise", for the purposes of our discussion. One light speed divides the set of possible fermions into two "superspecies", namely Normal and Dark fermions. The second light speed further divides each superspecies into four parts. The result is the eight fermion species in our universe (counting each quark as a different species). It includes both Normal and Dark sectors. It is also specifies the form of fermion species in the other HyperCosmos, and Second Kind HyperCosmos, spaces, all of which also have su(1, 1) multiple time coordinates.

The 16 fermions of the N = 7 Fundamental Reference Frame (our universe's) can be separated into a pair of eight subspecies if account is taken of the occurrence of each quark species as a triplet. This separation of fermion species leads to the structure of the Unified SuperStandard Theory (UST) and the Standard Model. It also leads to the form of Fundamental Reference Frames for all HyperCosmos (and 2nd Kind HyperCosmos) spaces.

The FRFs of higher r space-time dimensions have replicates of these 16 fundamental fermions. All fermions in the replicates are different with differing quantum numbers.

In the case of 2nd Kind HyperCosmos spaces we find their FRF's have multiples of only the sublight t_{01} Normal fermions. All the fermions in their higher space-time FRF replicates are different with differing quantum numbers. The Dark sector is missing in all 2nd HyperCosmos FRF's and spaces.

A complete, detailed spectrum of fundamental fermions in our universe, and other universes of all Cosmos Theory spaces, thus emerges, part of which appears in the UST and Standard Model for our universe.

t_{01}:	sublight **Normal**		superluminal **Dark**	
t_{02}:	subligh	superluminal	sublight	superluminal
	q-up-type and q-down-type	**e-type and v-type**	**q-up-type and q-down-type**	**e-type and v-type**
Number of fermions:	6	2	6	2

Figure 4.1. Separation of fundamental fermions in the N = 7 space's 16 dimension FRF. The superluminality of fermions is symbolic since the su(1, 1) coordinates are not space-time coordinates of a universe.

4.4 Scalar Bosons Corresponding to FRF Dimensions

We now consider the separation of scalar bosons including Higgs bosons when mapped from an FRF's dimensions. This case is particularly interesting because it evidences the separation effects of the two time coordinates which lead to sublight and superluminal particle differences.

Unlike the fermion case the irreducible representations of the symmetry groups have complex dimensions that may be separated into real and imaginary parts. The N = 7 space FRF maps to a set of 16 scalar bosons. These bosons are first separated by time coordinate t_{01} into Normal and Dark bosons. Then t_{02} separates the bosons in each sector into Higgs and non-Higgs bosons mirroring the fermion case above. See Fig. 4.2.

In the N = 7 space FRF Normal sector there is a non-Higgs set of three SU(3) bosons corresponding to the up-type and down-type quarks – totaling 6 bosons , and two Higgs bosons forming an SU(2) representation. The Dark sector has similar sets of bosons. See Fig. 9.2b.

t_{01}:	sublight **Normal**		superluminal **Dark**	
Bosons:	**8**		**8**	
t_{02}:	sublight	superluminal	sublight	superluminal
	Non-Higgs	**Higgs Bosons**	**Non-Higgs**	**Higgs Bosons**
Bosons:	**6**	**2**	**6**	**2**
Boson Group	SU(3)	SU(2)	SU(3)	SU(2)

Figure 4.2. Bosons separated by su(1, 1)'s two times coordinates – not space-time coordinates. There is one boson per fermion. Thus 6 bosons corresponding to the 6 quarks and two bosons corresponding to e and ν. Their group memberships are displayed as well. Bosons can be mapped from the FRF as well.

The N = 7 space FRF is mapped to a set of replicates in each FRF of higher space-time dimensions. Thus the set of split Fig. 4.2 bosons map to scalar bosons replicate sets. All bosons have separate quantum numbers and lay in different symmetry group representations.

4.5 Internal Symmetries of the HyperCosmos for N = 7

This section shows the N = 7 space FRF separation of internal symmetries due to the two time coordinates of su(1,1) – a group that appears in all HyperCosmos and all Second Kind HyperCosmos spaces.

t_{01}:	sublight Normal	superluminal Dark
	U(4)	U(4)
Real-valued Dimensions	8	8

t_{02}:	sublight	superluminal	sublight	superluminal
	6	2	6	2
	SU(3) Confinement	U(1) Higgs Mechanism	SU(3) Confinement	U(1) Higgs Mechanism
OR	4	4	4	4
	SO$^+$(1,3)	SU(2)⊗U(1)	SU(2)	SU(2)⊗U(1)
OR	4	4	4	4
	SU(2)	SU(2)⊗U(1)	SU(2)	SU(2)⊗U(1)

Figure 4.3. Split FRF dimensions mapped as real-valued representation dimensions to one of U(4)⊗U(4) for Generation and Layer symmetry groups, or [SU(3)⊗U(1)]^2 or SO$^+$(1,3)⊗SU(2)⊗[SU(2)⊗U(1)]^2 or [SU(2)⊗SU(2)⊗U(1)]^2 The SU(2) groups are Connection Groups. The SU(2)⊗U(1) groups are ElectroWeak Groups. See Fig. 2.6 for the N = 7 case.

4.5 Internal Symmetries of the HyperCosmos for N Greater than 7

The FRF's for higher space-time dimensions consist of replicates of the N = 7 dimensions. Thus the Symmetry Groups of these FRF's are replicates of the symmetry group maps in Fig. 4.3. The r = 6 FRF has two replicates (see eq. 3.8) of the N = 7 FRF. The symmetry groups of all replicates are different with different sets of quantum numbers. The replicates then map through GR transformations to the set of replicates embodied in the space's dimension array.

Comments

Note the sublight SU(3) group has a confining Strong interaction generated by a higher derivative interaction. Gravitation which should also be treated as sublight also has a higher derivative interaction that produces a MOND-like gravitational potential. See chapters 12 – 14 of Blaha (2022f) for details.

The split of particles and groups is based on the superluminal/subluminal separation of the coordinates of eq. 4.7 of the su(1, 1) group embedded within the creation/annihilation operators of fermion wave functions. It is not based on universe space-time coordinates.

The values of the nonzero dimensions of an FRF have no Physical meaning. The dimensions acquire Physical meaning only when mapped to particles and symmetry groups as above.

The values of the dimensions in the dimension array of a Cosmos space also have no Physical meaning. They acquire Physical meaning when mapped to particles and symmetry groups in universe creation.

The value of the dimension in a one[31] dimension FRF has no Physical meaning. Physical meaning is acquired after it is transformed to a space's dimension array and mapped to particles and symmetry groups in universe creation.

[31] One might call the dimension the "Adam" dimension since it is the source of all dimension array dimensions.

5. The Full HyperUnification Space

We now consider combined unification of all or some of the HyperCosmos spaces in a *Full HyperUnification Space* (a 42 space-time dimension space). Each space has a dimension array with d_{dN} elements. This array is expressed as a vector v_N with d_{dN} components residing in the space's unification space with space-time dimension r' given by eq. 3.1.

Now consider the sum of all ten HyperUnification spaces vectors:

$$v_S = \sum_{N=0}^{9} v_N \tag{5.1}$$

$$= 4^2 + 8^2 + 16^2 + 32^2 + 64^2 + 128^2 + 256^2 + 512^2 + 1024^2 + 2048^2$$

with

$$v_N = d_{dN} \tag{5.2}$$

These elements form a v_S-vector in the Full HyperUnification space. See Figs. 5.1 and 5.2. The sum of the ten vectors of the ten HyperCosmos HyperUnification spaces vectors is a vector d_{cS} of the Full HyperUnification Space:

$$d_{cS} = v_S = 5{,}592{,}400 \tag{5.3}$$

There is a corresponding set of square dimension array blocks whose total is

$$d_{dS} = 4^4 + 8^4 + 16^4 + 32^4 + 64^4 + 128^4 + 256^{4}\char`^ + 512^4 + 1024^4 + 2048^4$$
$$= 1.8765\times 10^{13} \tag{5.4}$$

The Full HyperUnification Space plays the role of unification for all ten HyperCosmos spaces. Each of the ten HyperCosmos HyperUnification space vectors of size d_{dN} in v_S individually has d_{dN} components. Correspondingly, each subvector has a $d_{dN} \times d_{dN}$ square array along the diagonal of a combined transformation which we call a *HyperUnification Transformation.* (Fig. 5.2)

Using eq. 3.1 we find the dimension array block defined by the d_{cS} vector is

$$d_H = 5{,}592{,}400^2 = 3.12749\times 10^{13} \tag{5.5}$$

If we treat this square array d_H as corresponding to a HyperCosmos space then we can calculate the corresponding space-time dimension

$$r_H = \log_2(d_H/16) \tag{5.6}$$
$$= 2\log_2(5{,}592{,}400) - 4 = 40.83$$

using

$$d_{dH} = 2^{22 - 2N_H} = 2^{r_H + 4} \tag{5.6}$$

The dimension r_H is non-integral. Noting that space-time dimensions are even numbered in HyperCosmos spaces. we define the *Full HyperUnification Space* to have

$$r_H = 42 \tag{5.7}$$

with

$$N_H = ½(18 - r_H) = -12$$

space-time dimensions.[32] It is the smallest space-time dimension that has a dimension array block containing the d_H dimension array block as shown in Fig. 5.2. (Note: the dimension array block d_{dN_H} has the same size as the General Relativistic transformation between the FRF and a static space-time.)

5.1 General Relativistic Transformation

The General Relativistic transformations of the r_H = 42 Full HyperUnification Space includes the General Relativistic transformations of any or all, HyperUnification spaces along its "diagonal" where blocks correspond to individual unification space General Relativistic transformations. See Fig. 5.1. It also includes much more in the form of arbitrary, not necessarily block diagonal, transformations that mix various HyperCosmos unification spaces.

The 42 space-time dimension array has the vector column size

$$d_{cN_H} = 2^{42/2+2} = 2^{23} = 8,388,608 \tag{5.8a}$$

with the number of array elements being

$$d_{dN_H} = 8,388,608^2 = 7.03687 \times 10^{13} \tag{5.8b}$$

Since d_{cN_H} is larger than d_{cS} there is an apparent excess in the 42 space-time dimension Full HyperUnification Space which we consider below. (See Fig. 5.2.)

We can define a set of HyperUnification Space General Relativistic transformations that transform in each of the individual HyperUnification sub-blocks separately. We can also define transformations that transform combinations of all 10 HyperUnification spaces including the entire set of HyperUnification spaces.

5.2 Full HyperUnification Space FRF

The FRF vector for the Full HyperUnification Space has d_{cN_H} components. Comparing it to the size of d_{cS} = 5,592,400-vector we find the General Relativistic transformations of the 10 HyperCosmos unification spaces are a subset of the 42 space-

[32] Negative values of N are allowed to specify Cayley number type dimension arrays. However they are not in the ProtoCosmos Model.

time dimension HyperUnification Space transformations. The column excess is 2,796,208 = 2,796,200 + 8 dimensions. The ratio of the larger box labeled A to the smaller box labeled B in Fig. 5.2 is exactly

$$d_{cN_H}/d_{cS} = 1.5 = 3/2 \tag{5.9}$$

Thus the B box column size d_{cB} satisfies

$$d_{cB} = \tfrac{1}{2}\, d_{cS} = 2{,}796{,}200 \tag{5.10}$$

and

$$d_{cB} = \tfrac{1}{2} \sum_{N=0}^{9} v_N \tag{5.11}$$

5.2.1 Second Kind HyperCosmos in 42 Dimension Space

The Full HyperUnification space holds the 10 HyperCosmos HyperUnification spaces plus the 10 Second Kind HyperCosmos HyperUnification spaces. The minimal space[33] that will hold the 10 HyperCosmos HyperUnification spaces has 42 space-time dimensions. (We require the Full HyperUnification space to have the same form as the HyperCosmos spaces but with a higher space-time dimension.)

The 42 space-time dimension Full HyperUnification space dimension array holds both the 10 HyperCosmos HyperUnification spaces and the 10 Second Kind HyperCosmos HyperUnification spaces dimension arrays (and the 42 space-time dimension General Relativistic transformations).

The Full HyperUnification space total dimensions expressed as a square array has 8,388,608 by 8,388,608 components. See Fig. 5.2. The 10 HyperCosmos HyperUnification spaces' dimension arrays occupy the 5,592,400 by 5,592,400 square array components block labeled by "A". The 10 Second Kind HyperCosmos HyperUnification spaces' dimension arrays occupy the 2,796,200 by 2,796,200 square array components block labeled by "B". The remaining 8 by 8 square array block labeled by "C" complete the Full HyperUnification space dimension array.[34]

For both A and B blocks, the HyperUnification space dimension blocks within them are strung along the diagonal as shown in Fig. 5.2.

Note that the length of block B, 2,796,200, is one half of the length of block A, 5,592,400. The difference is due to block A having both a 2,796,200 by 2,796,200 subblock for its Normal Matter sector and a 2,796,200 by 2,796,200 subblock for its Dark Matter sector. Block B only has a Normal Matter sector. Thus we see that the 42 space-time dimension array has three parts. This fact will be of interest later.

The 42 space exhibits a 1, 4, 9 sequence of blocks of dimensions. They are the squares of the 1, 2, 3 sequence of the A and B block sides and the total 42 space dimension array block side 2,796,200, 5,592,400, 8,388,608.

---

[33] A space with smaller than 42 space-time dimensions will not hold the dimension arrays of the 10 HyperCosmos HyperUnification spaces.

[34] See Blaha (2023b).

The B block holds the HyperUnification spaces of the 10 Second Kind HyperCosmos spaces that are described in Chapter 7.

5.2.2 Second Kind HyperCosmos Dark Matter Anyons?

The C block has 64 dimensions in an 8×8 square array with sides of 8 dimensions. This block corresponds to an r = 0 Second Kind HyperCosmos dimension array of size $2 \times 4 = 8$ dimensions. From eqs. 5.1 and 5.10 we see that there is an r = 0 term in block B. Thus block C duplicates that term.

We may view the r = 0 duplicate as a specification of a miniverse that could exist in our four dimension space. This miniverse could have eight[35] fermions in two sets of four Normal fermions (no Dark fermions) following the spectrum structure described earlier. They may each be SU(4) quark-lepton quadruplets[36]: one up-type; one down-type. Some of the possible symmetry groups are:

$$SU(4) \quad \text{or} \quad SU(2)\otimes SU(2)\otimes U(1) \quad \text{or} \quad SO^{+}(1, 3)\otimes SU(2)\otimes U(1)$$

If r = 0 miniverses exist in our universe in quantity then they could be candidates for Dark Matter. They only have the gravitational interaction outside of their miniverses. Since the space-time dimension is zero, their nature as fermions is clouded. They may support anticommutation relations; they may support commutation relations; they may support some intermediate form of commutation relations. Consequently these miniverses are a possible form of *anyon.*

The third possibility above opens the door to a space-time internal to a miniverse. The space-time of a PseudoFermion[37] wave function describes the PseudoFermion in a child space. The PseudoFermion's dimension array contains a set of dimensions for the internal space-time of a universe. The universe's internal space-time supports the wave functions of the particles and energies within the universe. The $SO^{+}(1, 3)$ group above is for the case of a miniverse with an internal space-time. In this case the minverse may have an internal dynamical evolution.

The absence of any experimental indication of Dark matter other than gravitation supports the possibility of an anyon Dark matter interpretation.

---

[35] The size of the miniverse dimension array.

[36] Possibly broken to $SU(3)\otimes U(1)$.

[37] Chapter 11 describes PseudoFermions and Independent PseudoFermions.

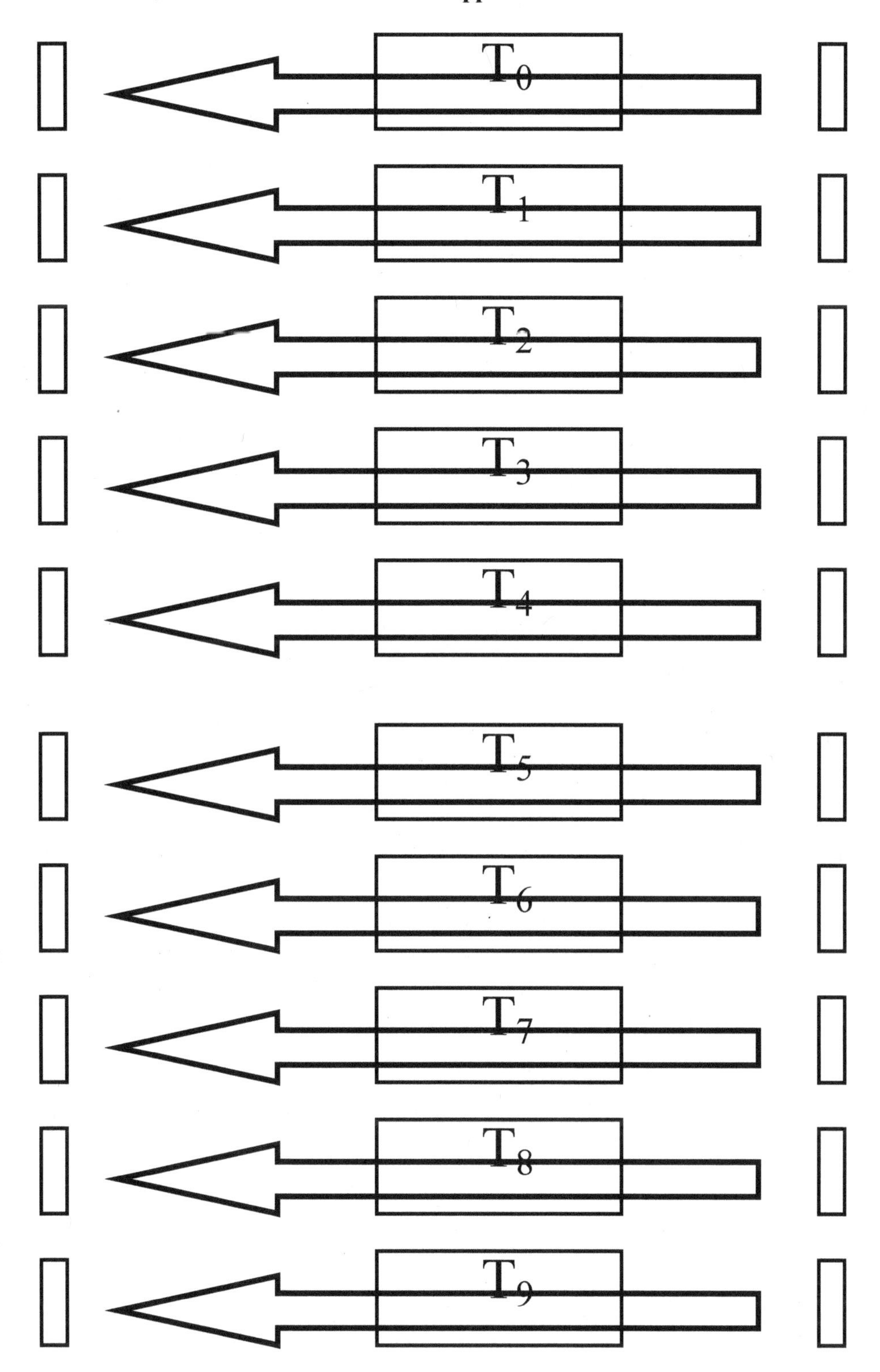

Figure 5.1. HyperCosmos FRF transformations in the Full HyperUnification Space.

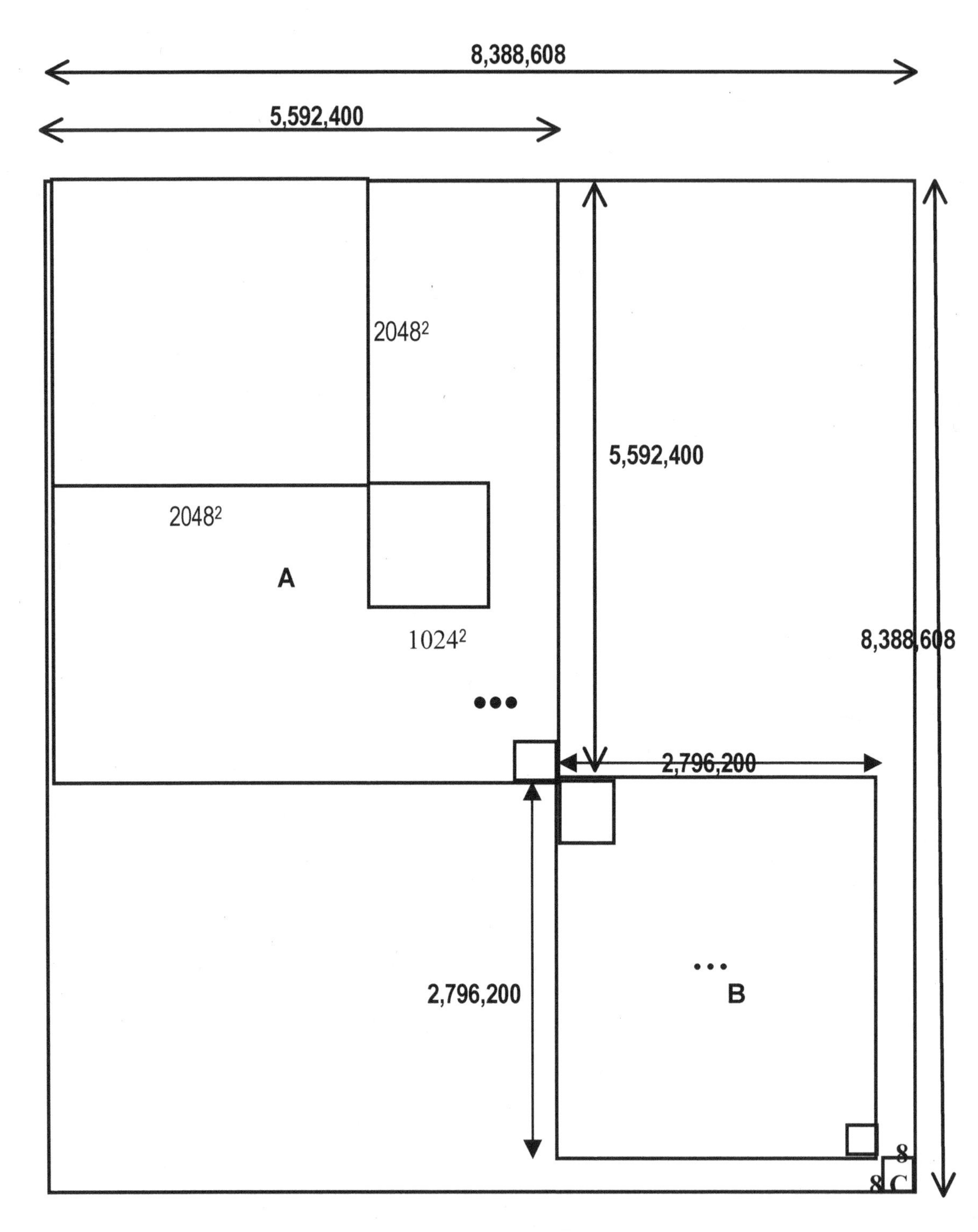

Figure 5.2. Form of the 42 space-time dimnesion of the Full HyperUnification space. All blocks are square. The figure is not drawn to scale.

6. One Dimension Generation in the Full HyperUnification Space and FRF

The FRF of a single HyperUnification space was reduced to one dimension in section 4.4. It is possible to reduce the combined FRF of the Full HyperUnification Space of Fig. 6.1 to one dimension.[38]

We demonstrate this possibility with a simple example. We reassemble the transformation to the form of Fig. 6.1 placing all space blocks of both the HyperCosmos and the Second Kind HyperCosmos to the extreme left.

Assuming the FRF has one non-zero dimension (element) a in row 1 of the Full HyperUnification Space FRF vector with all other components zero we define the vector

$$V = (a, 0, 0, \ldots) \tag{6.1}$$

and a *prototype* 42 dimension General Relativistic transformation

$$[T]_{ij} = b_i \ \delta_{i1} + (1 - \delta_{1i})(1 - \delta_{j1})c_{ij} \tag{6.2}$$

where $b_i \neq 0$ and the c_{ij} are numeric. Then T has entries b_i in column 1 and a matrix c_{ij} in the rows and columns for i, j = 2, 3, … . Consequently a full non-zero set of components d_{cN_H} is generated in the output vector which maps to the d_{cN_H} dimensions in vector V'. It is generated from one FRF dimension (element):

$$V' = TV \tag{6.3}$$

Thus we can generate the complete set of HyperCosmos and/or Second Kind HyperCosmos dimension arrays from a one dimension element in the 42 space-time dimension's FRF. The vector has non-zero components which become d_{cN_H} dimensions.

[38] We call the dimension the "Adam" dimension since it is the source of all dimension array dimensions.

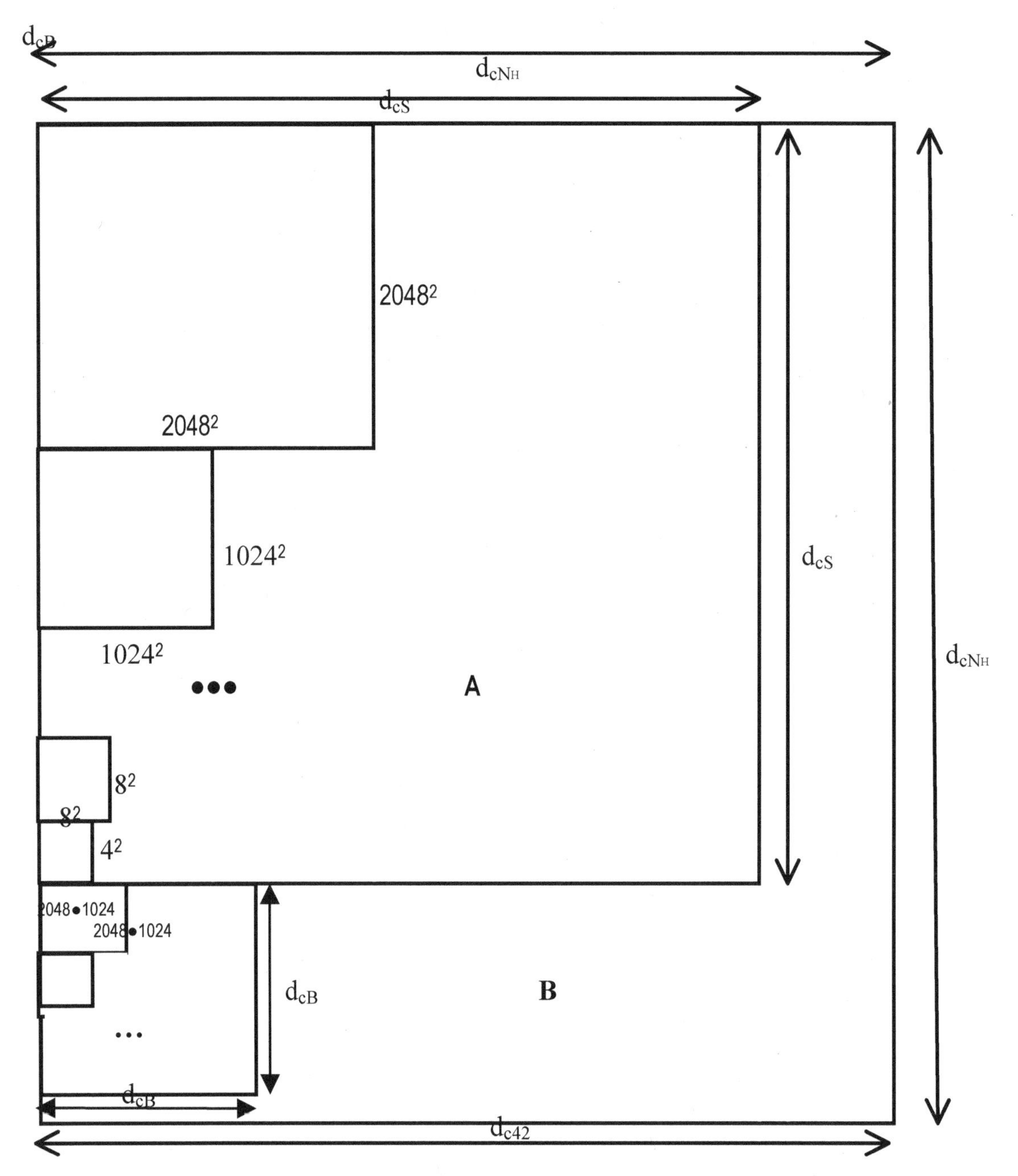

Figure 6.1. A HyperUnification transformation restructured to generate all dimentions from a one dimension FRF. The diagonal blocks are shifted to the left to support the transformation from a one dimension FRF of the 42 space-time dimension space. Block C is not shown This figure is not drawn to scale.

7. The HyperCosmos of the Second Kind

The form of d_{cR} for array B in chapter 5 suggests that the HyperCosmos of the Second Kind spectrum of spaces is similar to our original HyperCosmos with one basic change in the numerics:

$$d_{cR} = \frac{1}{2} \sum_{N=0}^{9} v_N \tag{7.1}$$

The vector v_N

$$v_N = d_{dN} \tag{7.2}$$

should be modified to

$$v_N' = \frac{1}{2} d_{dN} = d_{dN}' \tag{7.3}$$

which suggests

$$d_{dN}' = \frac{1}{2} 2^{r/2+2} 2^{r/2+2} = 2^{r/2+1} 2^{r/2+2} = 2^{r+3} \tag{7.4}$$

The Second Kind arrays are not square. They are rectangular.
There are several ways to obtain a $2^{r/2+1}$ factor:

1. Change the internal symmetry factor by a factor of two. This can be accomplished by eliminating the Dark sector: no Dark fermions and no Dark interactions. This would explain the existence of Dark matter without any interactions except gravity.

2. Remove the PseudoQuantum quantum field framework. This approach would remove the advantages of PseudoQuantum field theory.

3. Restrict General Relativistic (GR) transformations to b and $b^\dagger$ operators only. The d and $b^\dagger$ operators would not undergo GR transformations. These operators are half of the vector operators in GR transformations as evident in eq. 3.3 in chapter 3.

It appears that choice 1 is the best since it conforms to the known Dark matter features and best preserves a HyperCosmos formalism.

Thus a HyperCosmos of the Second Kind is possible. Whether Nature chooses it is an open question. We shall continue using the original HyperCosmos formalism.

The 42 space-time dimension space has a General Relativity that can mix the HyperCosmos spaces and the Second Kind HyperCosmos spaces. Thus there are gravitational interactions between them if the Second Kind set of spaces is instantiated with universes. We regard the HyperCosmos and the Second Kind HyperCosmos as siblings with the only interaction between them being gravitation at best.

The particles and symmetry groups of the Second Kind HyperCosmos spaces are not the same as the particles and symmetry groups in HyperCosmos spaces. The following figures show features of the N = 7 Second Kind HyperCosmos space and thus a Second Kind QUeST and Second Kind UST.

7.1 Two HyperCosmoses

We see that there are two HyperCosmoses: the original HyperCosmos and the HyperCosmos of the Second Kind. Either or both may be instantiated with universes. Our universe, at the moment, may be of either HyperCosmos. The key undecided question is the existence of a Dark sector.

The form of the Full HyperUnification Space supports two HyperCosmoses exactly. It provides significant support for the choice of a 42 dimension space-time.

Whether our universe is in the original HyperCosmos or in the Second Kind HyperCosmos is not known at present. The choice depends on the existence of the Dark sector.

7.2 Types of Transformations

The 42 space-time dimension Full HyperUnification space General Relativistic transformations are:

1. Transformations on the FRF of each HyperCosmos HyperUnification space individually.

2. Transformations on the FRF of each Second Kind HyperCosmos HyperUnification space individually.

3. Combined transformations on the FRFs of HyperCosmos HyperUnification spaces.

4. Combined transformations on the FRFs of Second Kind HyperCosmos HyperUnification spaces.

5. Transformations on the complete set of Full HyperUnification Space FRFs of both HyperCosmos and Second Kind HyperCosmos HyperUnification spaces.

7.3 Second Kind HyperUnification Spaces

Each of the ten Second Kind HyperCosmos spaces has a corresponding HyperUnification space. We assume the set of space-time dimensions of the Second Kind HyperCosmos is the same as the HyperCosmos. For space-time dimension r there are 2^{r+3} dimensions in the dimension array by eq. 7.4. Consequently the corresponding Second Kind HyperUnification space has vectors with 2^{r+3} components as indicated in Figs. 6.2 and 7.1.

Since the relation of the space-time dimensions is

$$2^{r+3} = 2^{r'/2+2} \tag{7.5}$$

for HyperCosmos-like spaces we find

$$r' = 2r + 2 \tag{7.6}$$

where r' is the space-time dimension of the HyperUnification space.

HYPERCOSMOS OF THE SECOND KIND SPACES SPECTRUM

Blaha Space Number $N = O_s$	Cayley-Dickson Number n	Cayley Number d_c	Dimension Array size d_d	Space-time-Dimension r	CASe Group $su(2^{r/2},2^{r/2})$ CASe
0	10	1024	1024 × 2048	18	su(512,512)
1	9	512	512 × 1024	16	su(256,256)
2	8	256	256 × 512	14	su(128,128)
3	7	128	128 × 256	12	su(64,64)
4	6	64	64 × 128	10	su(32,32)
5	5	32	32 × 64	8	su(16,16)
6	4	16	16 × 32	6	su(8,8)
7	**3**	**8**	**8 ×16**	**4**	**su(4,4)**
8	**2**	4	4 × 8	2	su(2,2)
9	**1**	2	2 × 4	0	su(1,1)

Figure 7.1. The HyperCosmos of the Second Kind space spectrum. The space for our universe, is number 7, with Cayley number 3 (which corresponds to octonions) is in bold type. Note the changed d_d column relative to the HyperCosmos.

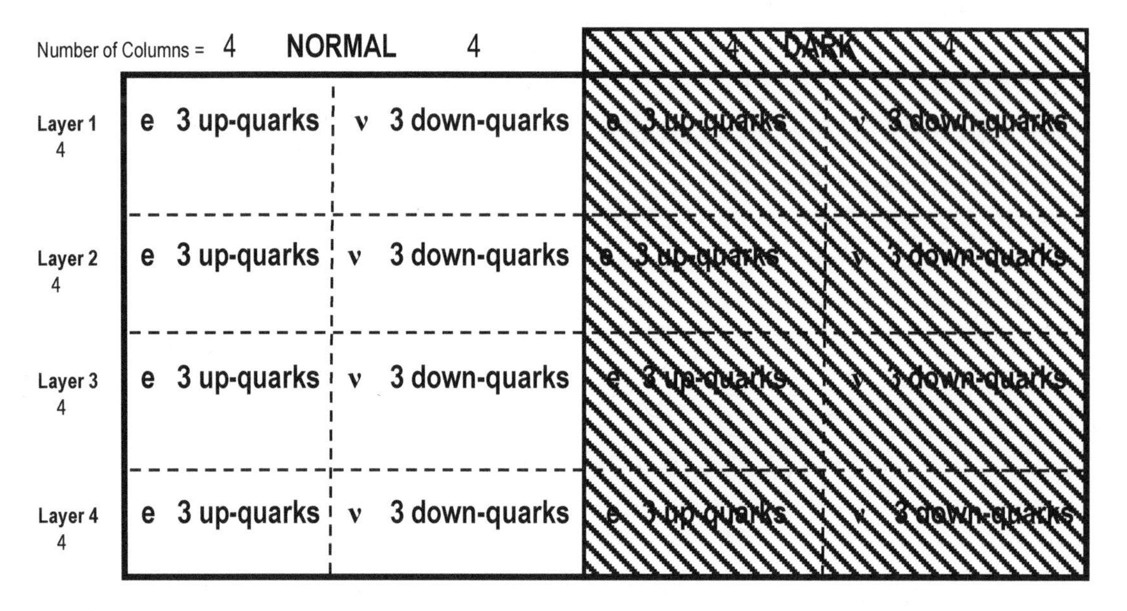

Figure 7.2. The HyperCosmos of the Second Kind 8×16 fermion spectrum tentatively arranged as SU(4)-plets that correspond directly with SU(4) (or SU(3)⊗U(1)) fermions. There are four layers. Each set of 4 fermions has 4 generations matching the number of rows in each layer. This Periodic Table is broken into Normal and Dark sectors. The absent Dark sector is indicated by the darkened part.

	NORMAL	DARK
Layer 1	U(4)⊗U(4) U(4)⊗U(4)	U(4)⊗U(4) U(4)⊗U(4)
Layer 2	U(4)⊗U(4) U(4)⊗U(4)	U(4)⊗U(4) U(4)⊗U(4)
Layer 3	U(4)⊗U(4) U(4)⊗U(4)	U(4)⊗U(4) U(4)⊗U(4)
Layer 4	U(4)⊗U(4) U(4)⊗U(4)	U(4)⊗U(4) U(4)⊗U(4)

Figure 7.3. The Second Kind "initial" distribution of sets of N = 7 symmetry groups. Each set is distinct and supports interactions only for the corresponding set of fermions (separately for Normal and Dark fermions). *Thus each set of 16 fermion generations has its own quantum numbers and interactions.* Each U(4)⊗U(4) set has a 16 real-valued dimension representation, which is of importance when we consider Fundamental Reference Frames. There is no Dark sector as indicated by the darkened part.

NORMAL

SU(3)⊗U(1) Generation U(4)	SU(2)⊗U(1)⊗SL(2, C) Layer U(4)
SU(3)⊗U(1) Generation U(4)	SU(2)⊗U(1)⊗U(2) Layer U(4)
SU(3)⊗U(1) Generation U(4)	SU(2)⊗U(1)⊗U(2) Layer U(4)
SU(3)⊗U(1) Generation U(4)	SU(2)⊗U(1)⊗U(2) Layer U(4)

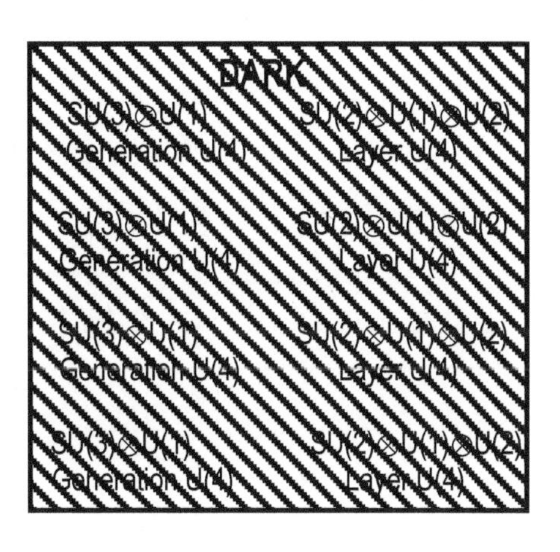

Figure 7.4. The transformed/broken sets of symmetries in N = 7 n Second Kind HyperCosmos space. Note each level has a 16 real dimension representation. This depiction is also evident in the Second Kind QUeST and UST. The SL(2, **C**) representation has four coordinates.[39] There is no Dark sector as indicated by the darkened part.

[39] The Lorentz Group $SO^{+}(1, 3)$ is often specified with an SL(2, **C**) representation.

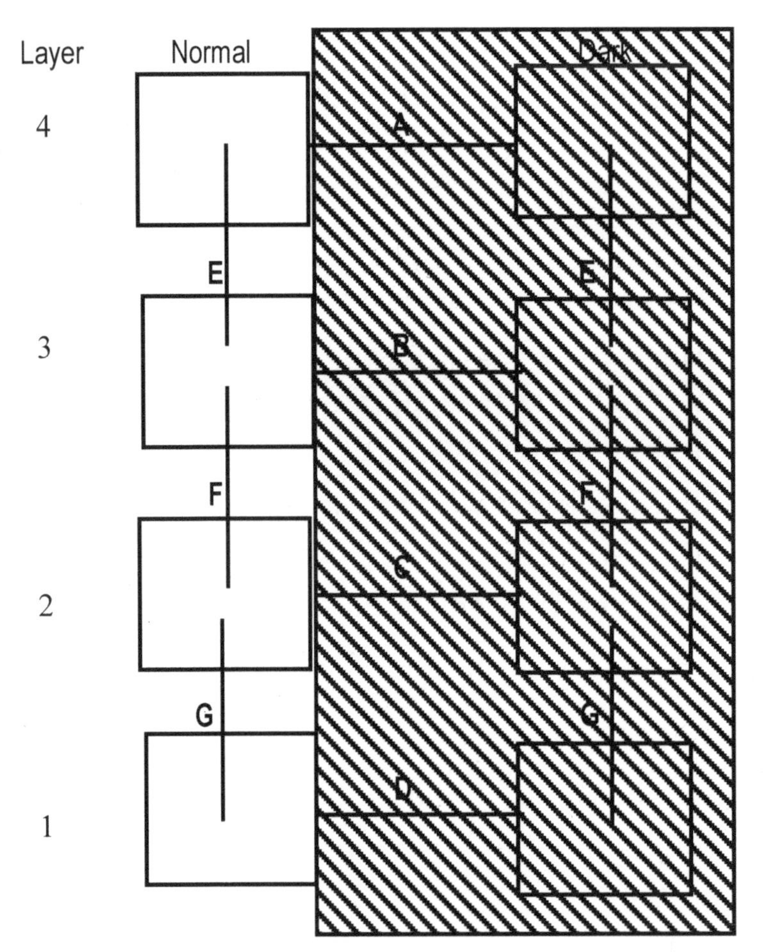

Figure 7.5. The three U(2) Connection groups[40] between the eight QUeST/UST blocks in the N = 7 Second Kind HyperCosmos. The Darkened part is not present in the Second Kind case. Connection groups are obtained by transfering 12 dimensions from QUeST space-time to internal symmetries with the consequent reduction of the space-time from four octonion (complex quaternion) coordinates to four real coordinates. The Connection groups generate rotations and interactions between corresponding fermions and vector bosons of each pair of blocks.

[40] Connection groups are discussed in Appendix 9-B.

8. The UltraUnification (UU) Space of the Full HyperUnification Space

The Full HyperUnification Space is a 42 space-time dimension space. It has a set of transformations of the form of Fig. 5.1. It also has a square dimension array of the same form as the transformation array: The number of rows and columns is d_{c42}. The dimension array has a similar content of dimension arrays of HyperUnification spaces of the HyperCosmos and the Second Kind HyperCosmos corresponding to the blocks of Fig. 5.2.

We can specify a set of unified transformations of these blocks in the same manner as we originally did for HyperCosmos spaces when we define HyperUnification spaces using

$$r' = 2r + 4 \tag{8.1}$$

with r = 42. The space-time dimension of its HyperUnification space, which we call the *UltraUnification (UU) Space*, is r' = 88. A 88 dimension General Relativistic transformation generally transforms all the contents of the Full HyperUnification Space. It therefore unifies all the space-time and Internal Symmetries within the set of HyperCosmos spaces and the Second Kind HyperCosmos spaces.

The UU space is needed to have an FRF with one non-zero dimension be transformed to the full set of dimensions in the 42 space-time dimension Full HyperUnification space and thence to the complete set of dimensions in the ten HyperCosmos spaces and in the ten Second Kind HyperCosmos spaces.

Note:

The defining feature of a HyperUnification space with space-time dimension r' of a space with space-time dimension r is that the number of components in the dimension r' GR transformation vector equals the total number of components d_{dN} *in the dimension array of the space of space-time dimension r.*

The 88 dimension UltraUnification space completes the Cosmos.[41] *The Cosmos has four levels as shown in Fig. 8.1. The total number of Cosmos spaces is*

!0 HyperCosmos spaces + their 10 HyperUnification spaces +
+ !0 Second Kind HyperCosmos spaces +
+ their 10 Second Kind HyperUnification spaces +

[41] The appearance of the dimensionless integers 42 and 88 in this Cosmos evokes the memory of the Eddington and Dirac (and others) conjectures on the appearance of 40 and 80 approximately in various combinations of universe parameters and physical constants. If some of their conjectures are valid they would be evidence for the author's Cosmos Theory. The dimensionless numbers 40 and 80 share the property of dimensionlessness with the dimensions of Cosmos Theory.

+ the 42 Dimension Full HyperUnification Space +
+ the 88 Dimension UU space
= 42 spaces

We thus find the interesting equality of the Full HyperUnification space dimension 42 and the total number of HyperCosmos spaces.[42] Whether this equality reflects a deeper underlying relation within the Cosmos' structure remains to be determined.[43] Fig. 8.1 displays the four levels and the 42 spaces of the complete author's *Cosmos Theory.*

The four levels of Fig. 8.1 may be viewed as forming a tetragram that reflects the structure of the Physical Reality (being) of the Cosmos.

8.1 Relation Between the Four Levels

The four levels of the Cosmos (Fig. 8.1) are intimately related:

A. The fourth level consists of the ten HyperCosmos spaces augmented by ten Second Kind HyperCosmos spaces.

B. The third level consists of 20 HyperUnification spaces. They are the HyperUnification spaces of the HyperCosmos and of the Second Kind HyperCosmos. The fourth level spaces each have a second level HyperUnification space. General Relativistic transformations of the HyperUnification space generate the dimension array of each r dimension HyperCosmos and Second Kind HyperCosmos space from a vector in the space's FRF. The vector may have only one non-zero dimension or may have a set of $2^{r/2+2}$ non-zero dimensions (or some other number of non-zero dimensions).

C. The second level is a 42 space-time dimension space with the same overall form as a HyperCosmos space. In this space the 20 HyperUnification spaces of the third level appear "along the diagonal."

D. The first level is the 88 space-time dimension UltraUnification space of the third level 42 dimension space. General Relativistic transformations of the UltraUnification space generate vectors yielding the dimension array of the 42 dimension third level HyperCosmos space from a vector in the fourth level space's FRF. The vector may have only one non-zero dimension or a set of 2^{42+4} $= 2^{46}$ non-zero dimensions (or some other number of non-zero dimensions).

---

[42] In addition there is an r = 0 space that we have suggested might be the space of anyon miniuniverses in our universe.

[43] The number 42 has a significance in a number in religious settings. We will not consider it in that context here.

A General Relativistic transformation of the 88 space-time dimension UltraUnification space cascades down to the fourth level HyperCosmos and Second HyperCosmos dimension arrays. Consequently it is possible to populate all HyperCosmos and Second Kind HyperCosmos dimension arrays with one non-zero dimension in the FRF vector of the 88 space-time dimension UltraUnification space.

8.2 Reduction of the UltraUnification Space FRF to One Dimension

We demonstrate this possibility with a simple example. Assuming the UU FRF has one non-zero dimension[44] (element) a in row 1 of the UU FRF vector with all other components zero we define the vector with d_{c88} components as

$$V = (a, 0, 0, \ldots) \tag{8.2}$$

and a *prototype* 88 space-time dimension General Relativistic transformation

$$[T]_{ij} = b_i \ \delta_{i1} + (1 - \delta_{1i})(1 - \delta_{j1})c_{ij} \tag{8.3}$$

where $b_i \neq 0$ and the c_{ij} are numeric. Then T has entries b_i in column 1 of all rows and a matrix c_{ij} in the rows and columns for i, j = 2, 3, … . Consequently a vector

$$V' = TV \tag{8.4}$$

is generated with all components non-zero. The generated vector V' has $d_{cN} = 2^{46}$ non-zero components which become d_{cN} dimensions. This vector can be written as a square array with 2^{46} components[45] for Blaha number[46] N = –35 using

$$N = 9 - r/2 \tag{8.5}$$

It is generated in a vector from one FRF dimension (element). Having the 42 space-time dimension array one can then map to the individual HyperCosmos and Second HyperCosmos dimension arrays.

We can generate the complete set of HyperCosmos and/or Second Kind HyperCosmos dimension arrays from a one dimension element.

---

[44] We call the dimension the “Adam” dimension since it is the source of all dimension array dimensions of all spaces.

[45] In agreement with eq. 6.8a.

[46] Negative values of N are allowed. They do not appear in the ProtoCosmos model.

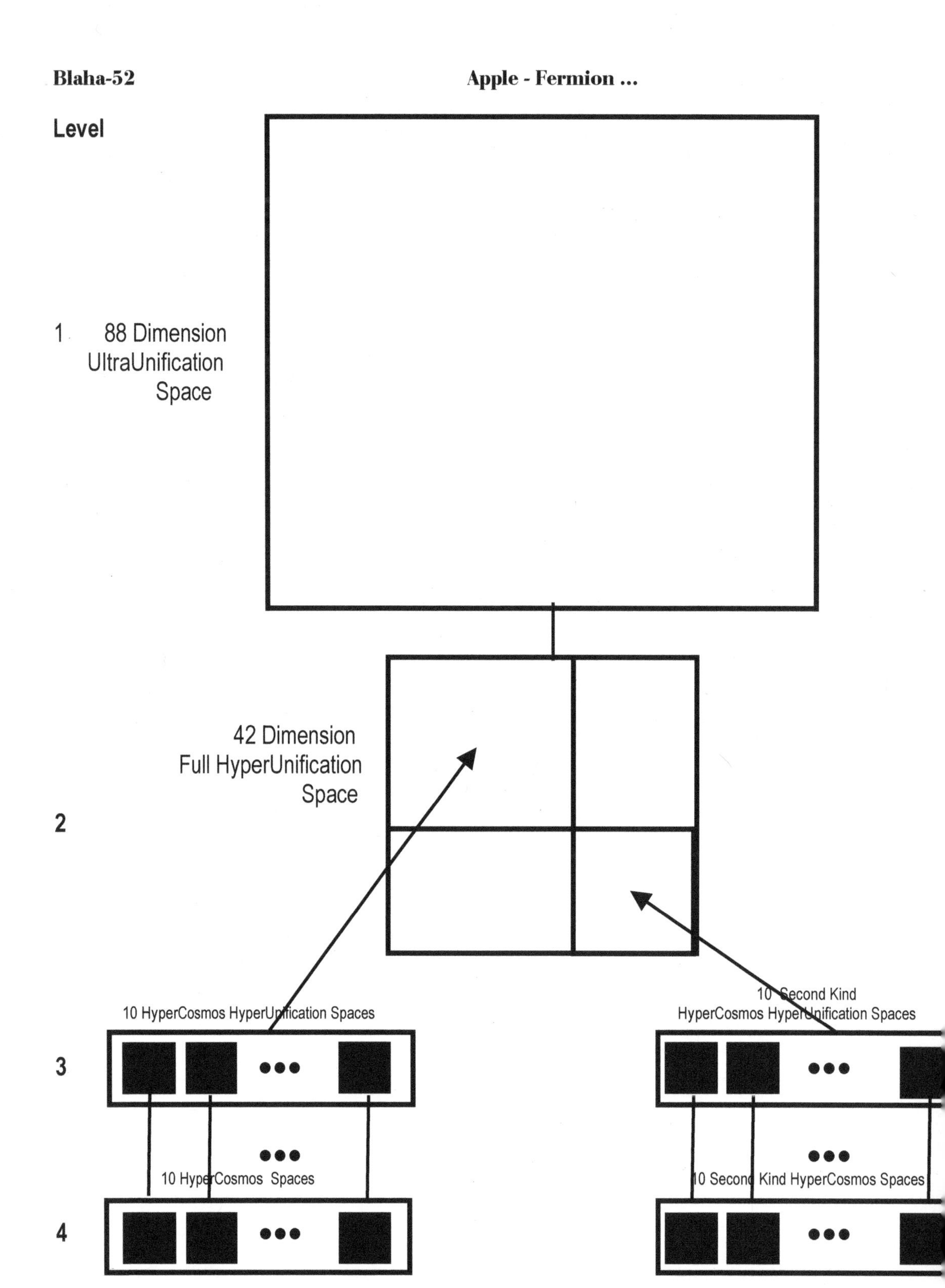

Figure 8.1. Diagram of the four levels of the Cosmos. They contain 42 spaces.

9. Unified SuperStandard Theory (UST) for HyperCosmos Spaces

The UST is the culmination of over two decades of effort by this author to develop a logical extension of The Standard Model of Elementary Particles.

The known part of the Standard Model contains three generations of fundamental fermions with each generation consisting of eight fermion species (e, ν, q_{1up}, q_{2up}, q_{3up}, q_{1down}, q_{2down}, q_{3down}). The Standard Model's known symmetry groups are SU(3)⊗SU(2)⊗U(1)⊗SO$^+$(1,3) – the strong interaction, ElectroWeak interaction, and space-time groups.

9.1 Our N = 7 Space UST

In the past decade we introduced the Unified SuperStandard Theory (UST) for our universe. It contained significantly more fundamental fermions and symmetry groups and a Dark sector that parallels the Normal sector of particles and fields. The preliminary UST is described in Blaha (2018e) and (2020a). In 2021-2 it added Connection Groups.

The features of the UST are:

1. 256 fundamental fermions appearing in four layers of four generations. 128 Normal fundamental fermions; 128 Dark fermions.[47] See Fig. 2.4.

2. A set of Symmetry Groups corresponding to the four layers – Normal and Dark sectors. See Fig. 2.6.

   $$[SU(3)\otimes U(1)\otimes SU(2)\otimes U(1)\otimes U(4)\otimes U(4)]^8\otimes SU(2)]^7\otimes SL(2, \mathbf{C})$$

   with SL(2, **C**) meaning SO$^+$(1, 3) where $[SU(2)]^7$ represents the Connection Groups. (See Appendix 9-B.) The U(4)⊗U(4) factors represent the Generation and Layer Groups. See Appendix 9-A. Each symmetry group is fully independent of other groups. Thus each of the four levels has a separate set of symmetry groups.

3. The UST has Higgs bosoms and other scalar bosons (See Chapter 4.)

4. The UST is *totally* renormalizable in perturbation theory by using the author's Two Tier Renormalization Program first presented in Blaha (2002) and (2005a). Two Tier renormalization also makes higher space-time dimension Quantum Field Theories fully renormalizable unlike other renormalization programs.

---

[47] We assume the N = 7 HyperCosmos space. If our universe is based on the Secônd Kind HyperCosmos space than the corresponding UST would omit the Dark sector.

5. The Dark sector has its own set of symmetry groups which exactly parallel that of the Normal sector.

6. The seven Connection Groups connect layers in the Normal and Dark sectors and also connect the Normal and Dark sectors. See section 9-B and Fig. 9-B.2.

Chapter 10 discusses UST symmetry groups, representations and interactions.

9.1.1 Comments

The UST fermion spectrum contains the known three generations of fermions. Parts of the spectrum have not been found as yet due to their large masses. The Dark sector has not been found experimentally. See Fig. 2.4.

The UST symmetry groups have also been only partially found. The symmetry groups that have not been found as yet presumably have very large masses. Symmetry groups are listed in Fig. 9.1. The seven Connection Groups (not shown) have also not been found as yet.

The Normal and Dark sectors are distinguished by an additive Darkness quantum number D with the value 1 for Normal sector particles and symmetry group gauge fields and -1 for Dark sector particles and symmetry group gauge fields.

The UST is fully united within the N = 7 HyperUnification space formalism based on 12 space-time dimension GR transformations which was discussed earlier. The values of couplings constants are thus not relevant.

It is important to note that the structure of the UST symmetry groups (Fig. 9.1) follows directly from the structure of fermion creation/annihilation operators (section 2.9) together with the structure implied by su(1, 1) (which also follows from creation/annihilation operator considerations) as shown in chapter 4. This structuring is modified to some extent by spontaneous breakdowns.

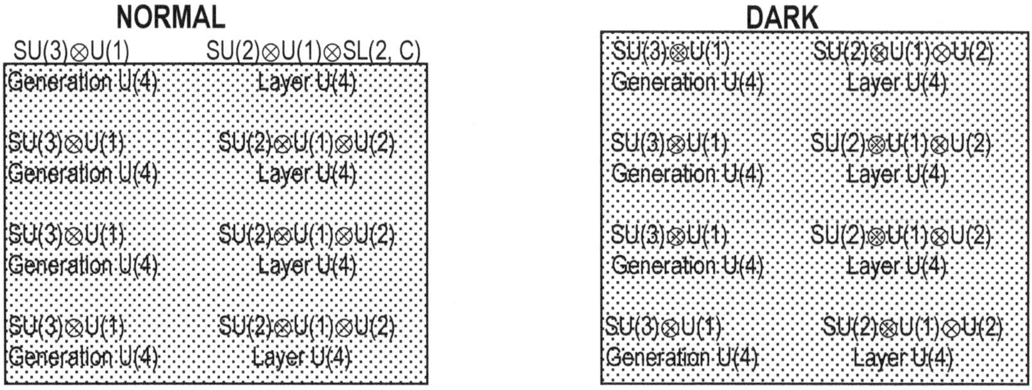

NORMAL

$SU(3)\otimes U(1)$ $SU(2)\otimes U(1)\otimes SL(2, C)$
Generation U(4) Layer U(4)

$SU(3)\otimes U(1)$ $SU(2)\otimes U(1)\otimes U(2)$
Generation U(4) Layer U(4)

$SU(3)\otimes U(1)$ $SU(2)\otimes U(1)\otimes U(2)$
Generation U(4) Layer U(4)

$SU(3)\otimes U(1)$ $SU(2)\otimes U(1)\otimes U(2)$
Generation U(4) Layer U(4)

DARK

$SU(3)\otimes U(1)$ $SU(2)\otimes U(1)\otimes U(2)$
Generation U(4) Layer U(4)

$SU(3)\otimes U(1)$ $SU(2)\otimes U(1)\otimes U(2)$
Generation U(4) Layer U(4)

$SU(3)\otimes U(1)$ $SU(2)\otimes U(1)\otimes U(2)$
Generation U(4) Layer U(4)

$SU(3)\otimes U(1)$ $SU(2)\otimes U(1)\otimes U(2)$
Generation U(4) Layer U(4)

Figure 9.1. The transformed/broken sets of symmetries in UST (and QUeST) and in N = 7 HyperCosmos space.. The darkened parts have not as yet been found. Note each element has a 16 real dimension representation. The SL(2, **C**) representation has four coordinates.[48]

9.2 Possible Second Kind UST

It is possible that our universe does not have a Dark matter sector. Then the space of our universe would be a Second Kind HyperCosmos space. This type of space has the same Normal sector as HyperCosmos spaces. See chapter 7.

9.3 Higher Space-Time Dimension (N < 7) Space's UST "Equivalent"

The higher space-time dimension spaces (N < 7) have generalizations of our N = 7 space's UST. The basis of these Higher UST's, which we will denote as HUST's, is the quadrupling of our UST as space-time dimensions r increase by twos. This is evident in the pattern of dimension array sizes in the HyperCosmos space spectrum.

Thus we may view the r = 6 space (for a Multiverse universe) HUST as the quadruple of our UST with 1024 fermions and 1024 symmetry group irreducible representations. Figs. 2.4 and 9.1 are "quadrupled."

The r = 8 space HUST and universe is quadruple the r = 6 space HUST. And so on.

The only major point of difference in HUST structure is in the Connection groups. The differences are considered in Appendix 9-B. See Figs. 9-B.3, 9-B.4, and 9-B.5 and accompanying text for the r =4, 6, and 8 Connection Group cases.

Note all HUSTs have fully renormalizable perturbation theories due to our Two Tier quantum theory formalism.

It is important to note that the structure of each HUST set of symmetry groups follows directly from the structure of the r space-time dimension[49] fermion creation/annihilation operators (section 2.9) together with the structure implied by su(1, 1) (which also follows from creation/annihilation operator considerations) as shown in chapter 4. This structuring is again modified to some extent by spontaneous breakdowns.

HyperCosmos space structures, and thus HUSTs, have both a Normal and Dark sector. Second Kind HyperCosmos spaces (and HUSTs) have only a Normal sector for any space-time dimension r. Thus a 2nd Kind HUST has half the fermions and half the irreducible symmetry group dimensions as the corresponding HyperCosmos HUST.

---

[48] The Lorentz Group $SO^+(1, 3)$ is often specified with an SL(2, **C**) representation.

[49] The space-time dimension determines the spins and the number of creation/annihilation operators and thus the HUST dimension array structure.

Appendix 9-A. Generation and Layer Groups

9-A.1 U(4) Generation Groups

In the Big Bang all particles were massless and all symmetries unbroken. Hence the four Normal particle number symmetries, and the four Dark particle number symmetries, are all "conserved" in the Big Bang. Afterwards conservation laws are then broken.

We define two particle number operators for normal up-quark particles and down-quark particles, B_{uq} and B_{dq}. Similarly we define two particle number operators for normal species "e" (electron) particles and species "ν" particles, B_e and $B_ν$. Similarly we define Dark matter equivalents:[50] B_{De}, $B_{Dν}$, B_{Duq}, and B_{Ddq}.

In the absence of symmetry breaking these fermion particle number operators would be conserved. Thus there are two sets of "diagonal" operators with associated U(4) groups for the Normal and Dark sectors. They are part of the Normal U(4) Generation Group and the Dark U(4) Generation Group.

The fermion fundamental representation of a U(4) group has four fermions. U(4) has rotations, and also interactions of the form $\overline{\Psi}\gamma\cdot B\cdot T\Psi$ where Ψ is a fermion four-vector, B is a 16 component U(4) gauge field, and T consists of 16 component 4 × 4 U(4) arrays.

In the case of the Generation Group the gauge fields have electric charge zero. Since the four species have different electric charges (1, 0, 2/3, -1/3) the U(4) gauge boson fields cannot mix the fermions of different species. Generation Group interactions are diagonal[51] in fermion species (e, ν, up-quark, and down-quark species).

Consequently the U(4) Generation Group for each layer[52] must have a reducible representation D consisting of a set of four fundamental U(4) representations, D_e, $D_ν$, D_{upq}, and D_{dnq}, appearing in blocks along the diagonal of D. Each block is a separate U(4) irreducible representation for a species (due to the electric charge superselection rule.) There is a U(4) Generation group for each of the four layers of the Normal and Dark sectors totaling to 8 generation groups.

There are four generations of each species in the Normal and in the Dark matter sectors. The four generations for each fermion species: e, ν, up-quark, and down-quark each furnish a U(4) fundamental representation within the reducible representation D. The fourth generation of normal fermions has not as yet been found due to their extremely large masses.

---

[50] By analogy, we assume that there are four species of Dark matter: charged Dark leptons, neutral Dark leptons, Dark up-type quarks, and Dark down-type quarks. Thus we are led to the Dark particle numbers: Dark Baryon Numbers, and Dark Lepton Numbers shown above.

[51] ElectroWeak interactions can cross between species due to their charged gauge vector bosons.

[52] For the Normal and Dark sectors separately.

The Generation Group rotates the fundamental fermions of each fundamental representation separately for each of the four species of each of the four layers.[53] Thus the Generation Group guarantees that all generations of each species have the same electric charge and other quantum numbers.

The U(4) Generation Group also specifies a gauge field interaction among the fermions of its fundamental representation, species by species, for both Normal and Dark sectors. The form of the interactions for the Normal sector for each fermion layer is:

$$g_e\overline{\Psi}_e\gamma\cdot B_e\cdot T\Psi_e + g_v\overline{\Psi}_v\gamma\cdot B_v\cdot T\Psi_v + g_{upq}\overline{\Psi}_{upq}\gamma\cdot B_{upq}\cdot T\Psi_{upq} + g_{dnq}\overline{\Psi}_{dnq}\gamma\cdot B_{dnq}\cdot T\Psi_{dnq} \tag{9-A.1}$$

where g_e ... are coupling constants, the gauge vector fields are B_e ... , and the Ψ_e ... are 4-vectors of fermions of the four generations of each species in a layer.

The gauge vector bosons of the Generation Group have large masses. If the conservation of the fermion particle numbers is broken then we view it as a consequence of Generation Group symmetry breaking.

Generation Group rotations guarantee the internal quantum numbers of each generation of each species are the same since symmetry breakdown is not present at the instant of the Big Bang.

The above discussion applies similarly to the Dark sector. Thus there are 8 Generation Groups in total.

9-A.2 U(4) Layer Groups

The set[54] of particle number operators can be extended if we take account of the fourfold fermion generations.

We can subdivide the above particle number sets into four additional particle numbers *per generation*. For the ith generation (of the four generations) we define

L_{ie} – The "e" species particle number for the ith generation
L_{iv} – The ν species particle number for the ith generation
L_{iuq} – The up-quark species particle number for the ith generation
L_{idq} – The down-quark species particle number for the ith generation

L_{iDe} – The Dark "e" species particle number for the ith generation
L_{iDv} – The Dark ν species particle number for the ith generation
L_{iDuq} – The Dark up-quark species particle number for the ith generation
L_{iDdq} – Dark down-quark species particle number for the ith generation

for each generation i = 1, 2, 3, 4. Individual fermions have positive $L_{ia} = +1$ values and antifermions have negative $L_{ia} = -1$ values for each species.

---

[53] There are separate Generation groups for each layer.

[54] Here again, in the Big Bang all particles were massless and all symmetries unbroken. Hence particle numbers are "conserved" in the Big Bang. Conservation is then broken afterwards in most cases.

At this point we have a set of four particle number operators for each of four generations (i = 1, 2, 3, 4) of fermions in the Normal sector and similarly in the Dark sector. We then define a U(4) group framework for each set of particle numbers.

The only way to specify fundamental representations for each of the four sets in a sector is to assume there are four layers, with each layer having four generations, and with a fundamental U(4) representation defined for each generation composed of fermions from each layer. Thus there are four Layer Groups for each Normal and each Dark sector: a Layer Group for generation 1, a Layer Group for generation 2, and so on.

The Layer Groups are also "split" by species due to the electric charge superselection rule. Each Layer Group is diagonal in the four fermion species. All their gauge fields are electrically neutral. Thus the Layer Group fundamental representations of the Normal sector total 16.[55] The Layer Group fundamental representations of the Dark sector also total 16. There are 4 Normal Layer Groups and 4 Dark Layer Groups.

Consequently each of the four U(4) Layer Groups in the Normal fermion sector has a reducible U(4) representation D_j for j = 1, 2, 3, 4. Each reducible representation is composed of four irreducible U(4) representations for each species due to the electric charge superselection rule:

$$D_j = D_{je} + D_{j\nu} + D_{jupq} + D_{jdnq},$$

for j = 1, 2, 3, 4.

There are four layers of each species in the Normal and in the Dark matter sectors. The second, third and fourth layers of normal fermions has not as yet been found due to their extremely large masses.

A Layer Group rotates the fundamental fermions of each fundamental representation separately for each of the four species of each of the four generations.

The Layer Groups guarantee that all layers of each species have the same electric charge and other quantum numbers.

Each U(4) Layer Group also specifies a gauge field interaction among the fermions of its fundamental representation, species by species, for both Normal and Dark sectors. The form of the interactions is:

$$g_{ei}\overline{\Psi}_{ei}\gamma\cdot C_{ei}\cdot T\Psi_{ei} + g_{vi}\overline{\Psi}_{vi}\gamma\cdot C_{vi}\cdot T\Psi_{vi} + g_{upqi}\overline{\Psi}_{upqi}\gamma\cdot C_{upqi}\cdot T\Psi_{upqi} + g_{dnqi}\overline{\Psi}_{dnqi}\gamma\cdot C_{dnqi}\cdot T\Psi_{dnqi} \tag{9-A.2}$$

for i = generation = 1, … , 4, where g_{ei} … are coupling constants, the gauge fields are C_{ei} … , and the Ψ_{ei} … are 4-vectors of fermions formed of the ith generation fermions in each layer of each species.

The gauge vector bosons of the Layer Groups also have large masses. If the conservation of the fermion particle numbers is broken then we view it as a consequence of Layer Groups symmetry breaking.

---

[55] Four Layers groups irreducible representations for the generations × four species = 16 irreducible representations.

Layer Group rotations guarantee the internal quantum numbers of each layer of each species are the same since symmetry breakdown is not present at the instant of the Big Bang.

The above discussion applies similarly to the Dark sector. There are 16 Dark Layer Groups.

Fig. 7.3 shows the fundamental fermion spectrum with the representations of the Generation groups and Layer groups indicated.

Experimentally, we know of three generations of fermions—the lowest 3 generations of the lowest level. The remaining 4th generation and three layers of fermions are of much higher mass and are yet to be found.

See Blaha (2019g) and (2018e) for a detailed discussion of the Layer Groups. We note in passing that the symmetries of these number operators are badly broken. Yet the underlying group structure remains.

Appendix 9-B. Connection Group Symmetries and HyperCosmos Space-Time Coordinates

Hypercomplex numbers are known to be related to symmetry groups. In this section we consider the multiple space-time symmetries that appear in the separation of dimension arrays into representations of groups. We have suggested in Blaha (2021b), (2021e) and (2021g) that the set of hypercomplex space-time dimensions be transformed to a set symmetry groups in each HyperCosmos space.

There is a two-fold justification for this procedure: 1). There is no evidence for the existence of hypercomplex numbers in our universe. 2) The generated set of symmetry groups has the important purpose of providing ultra-weak interactions uniting Normal and Dark matter. Without unification, Dark matter becomes physically irrelevant for elementary particle physics.

We begin by noting the space-time dimension is set by the dimension array:

$$r = \log_2 (d_{dN}/16) \qquad (9\text{-B.1})$$

For N = 9 the space-time dimension is 0 since $d_{dN} = 16$. For N = 7 (our universe's space) the space-time dimension is 4 since $d_{dN} = 256$. We will consider the cases of QUeST, UTMOST, and Maxiverse to illustrate the method.

9-B.1 Hypercomplex Coordinates Transformed to Symmetry Groups in Our Universe N = 7

Our QUeST formulation for our N = 7 universe has a $d_{d7} = 256$ component dimension array. A preliminary view of the symmetry groups that it implies appears in Fig. 9-B.1. Note that there are 32 dimensions that are initially allocated to space-time dimensions suggesting a hypercomplex space-time. We chose to allocate 4 real dimensions to obtain the 4-dimension space-time implied by eq. 9-B.1. The remaining 28 real dimensions we allocated to additional symmetry groups—namely seven U(2) groups. We call these groups Connection groups since their role is to "connect" fermions residing in different fermion spectrum blocks.

The structure of the seven additional U(2) groups is not specified. We chose to use a reasonable physical principle to allocate them. We believe their role is to "connect" fermions in different blocks since the fermions within each fermion block have "known" interactions. Note that there are initially eight blocks, each with their own set of symmetry groups and corresponding interactions, and initially no interactions between the eight blocks. If there are no block interactions (except gravity), then the Physics of the fermion set is conceptually disjoint. Thus we choose to implement inter-block interactions as in Fig. 9-B.2 following the principle:

A fermion in any block has interactions either directly, or indirectly, with every other fermion in every other block.

Fig. 9-B.2 shows an implementation of this principle. The horizontal lines in Fig. 9-B.2 indicate 1:1 transformations between all corresponding fermions of each "Normal" and "Dark" block. The three "angled" lines indicate 1:1 transformations between corresponding fermions of a "Normal" and a "Dark" fermion block in the layer above it. The result is a see-saw type of pattern.

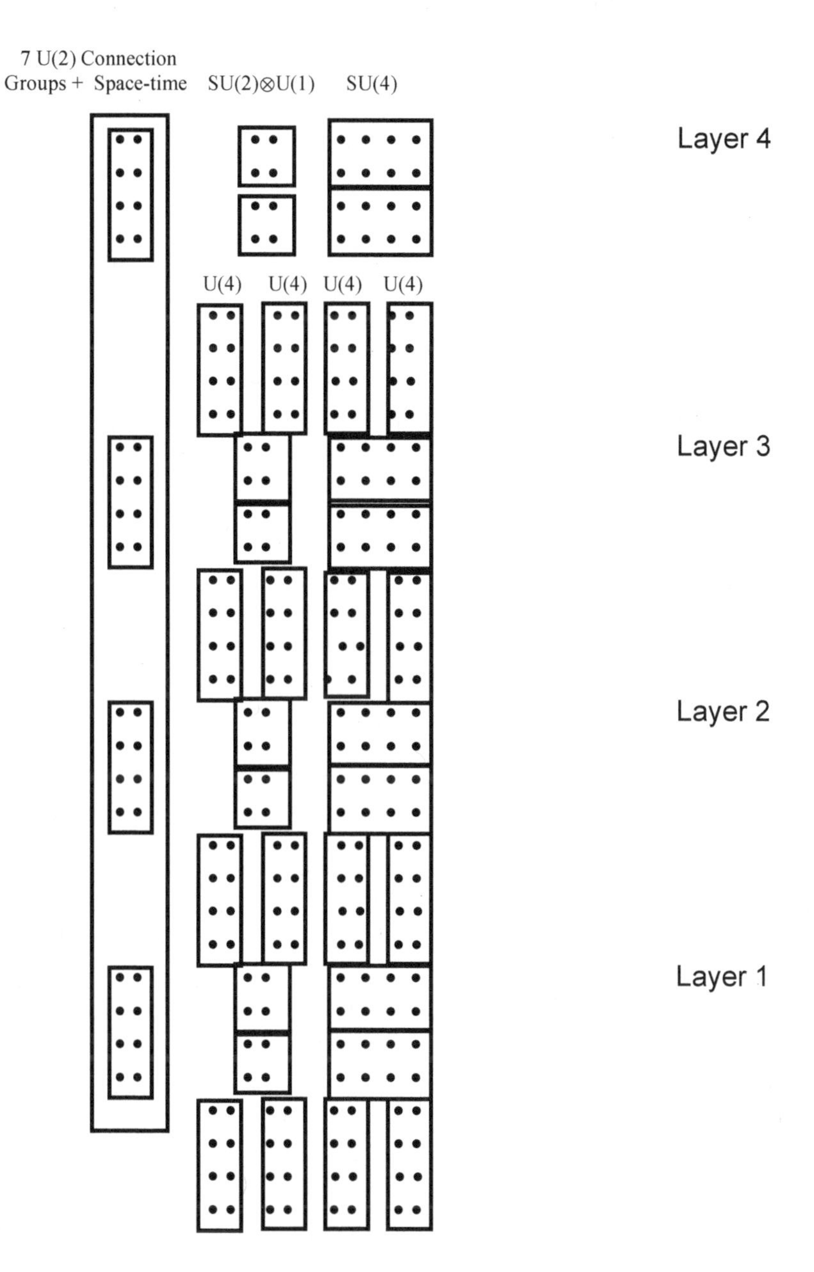

Figure 9-B.1. The four layers of QUeST, UST and QUeST internal symmetry groups (and space-time) with SU(4) before breakdown to SU(3)⊗U(1). Note the left column of blocks are combined below to specify a 4 dimension real space-time plus seven U(2) Connection groups. Note each layer has 64 dimensions = 56 + 8 dimensions.

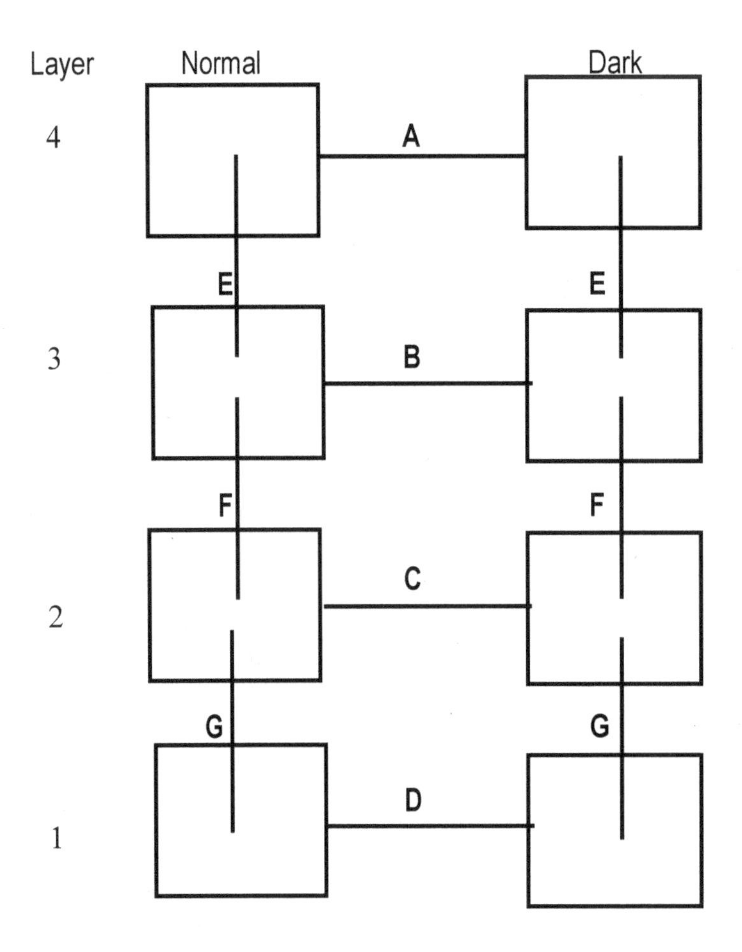

Figure 9-B.2. The seven U(2) Connection groups (shown as 10 lines) between the eight QUeST/UST blocks. Connection groups are obtained by transfering 28 dimensions from QUeST space-time to internal symmetries with the consequent reduction of the space-time from four octonion (complex quaternion) coordinates to four real coordinates. The Connection groups generate rotations and interactions between corresponding fermions and vector bosons of each pair of blocks. The Normal and Dark sector U(2) vertical connections above (E, F, G) represent the same U(2) groups.

9-B.1.1 The U(2) Connection Groups

The seven U(2) Connection groups of Fig. 9-B.2 generate "rotations" and interactions between *corresponding* fermions and vector bosons of each pair of blocks of the eight blocks of fermions in QUeST/UST.

9-B.1.1.1 Horizontal Lines

The horizontal lines in Fig. 9-B.2 (A, B, C, and D) each represent a U(2) Connection group that "rotates" two *corresponding* fermions in the Normal and Dark sectors of each layer. Thus a Normal e is "rotated" with a corresponding Dark e, and so on.

Each of the four horizontal Connection Groups has a reducible U(2) representation D that is the sum[56] of 4*8 = 32 irreducible U(2) representations. We may view the irreducible representations D_j as strung along the diagonal.

$$D = \sum_{j=1}^{32} D_j \tag{9-B.2}$$

for each of the U(2) groups of the four horizontal lines in Fig. 9-B.2.

The U(2) group also specifies gauge field interactions between corresponding fermions in each layer of the Normal and Dark sectors of the form

$$g\overline{\Psi}_{Nn}\gamma\cdot A\cdot T\Psi_{Dn} \tag{9-B.3}$$

where N indicates a Normal fermion and D indicates the corresponding Dark fermion, with A being a U(2) gauge vector boson, and n the label for corresponding fermions.

These U(2) transformations imply that the Normal and Dark sectors have the same species.

9-B.1.1.2 Vertical Lines

The pairs of vertical lines in Fig. 9-B.2 (E, F, G) each represent a U(2) Connection group that "rotates" sets of two *corresponding* fermions in adjacent layers as shown in Fig. 9-B.2 in the Normal and Dark sectors. Thus a Normal e in layer 1 is rotated with a corresponding Normal e in layer 2, and so on.

Each of the three (six counting both Normal and Dark lines in Fig. 9-B.2) vertical Connection Groups has a reducible U(2) representation D that is the sum of 64 irreducible U(2) representations.[57] We may view D as an array of 64 U(2) irreducible representation dimensions D_j strung along the diagonal.

$$D = \sum_{j=1}^{64} D_j \tag{9-B.4}$$

for each of the U(2) groups of the 3 (6) horizontal lines in Fig. 9-B.2. Note the 64 irreducible representations include both Normal and Dark sectors of a layer. [58]

Each U(2) group also specifies a gauge field interaction between corresponding fermions in adjacent layers for both Normal and Dark sectors:

---

[56] Eight fermions per generation ▪ four generations, thus accounting for each fermion in a block.

[57] Eight fermions per generations ▪ four generations ▪ 2 types of matter (Normal and Dark).

[58] There are 64 fermions in total for each of the four layers of NEWQUeST/NEWEST.

$$g\overline{\Psi}_{nl_1}\gamma\cdot A\cdot T\Psi_{nl_2} \qquad (9\text{-}B.5)$$

where l_1 and l_2 designate layers, A is a gauge field vector boson, and n the label for corresponding fermions.

Each E, F, and G U(2) group reducible representation includes both Normal and Dark sectors.

9-B.1.2 The Connection Groups are UltraWeak Interactions

Since there is no convincing experimental evidence for particle interactions between Normal and Dark matter, or between Normal fermion layers the Connection groups appear to be UltraWeak.

9-B.2 UTMOST with Six Real Space-Time Coordinates (Dimensions)

The UTMOST space is a Blaha number N = 6 HyperCosmos space. It is the Megaverse (Multiverse) with 6 real-valued space-time coordinates by eq.9-B.1 above. It has a d_{d6} = 1024 dimension array.

We view the UTMOST dimension array as composed of four copies of the N = 7 dimension array. Initially 4·32 = 128 dimensions are for space-time coordinates.

We allocate 4·28 = 112 dimensions to each of the four QUeST copies within UTMOST to give each copy 7 SU(2) Connection groups. Therefore 4* 28 = 112 of the 128 space-time dimensions of UTMOST are mapped. The remaining 16 dimensions give 6 real-valued space-time dimensions to the Megaverse space, 8 dimensions to a Megaverse SU(4) Connection group, and 2 dimensions to a U(1) group for all fermions in the Megaverse. See Figs. 9-B.3 and 9-B.4.

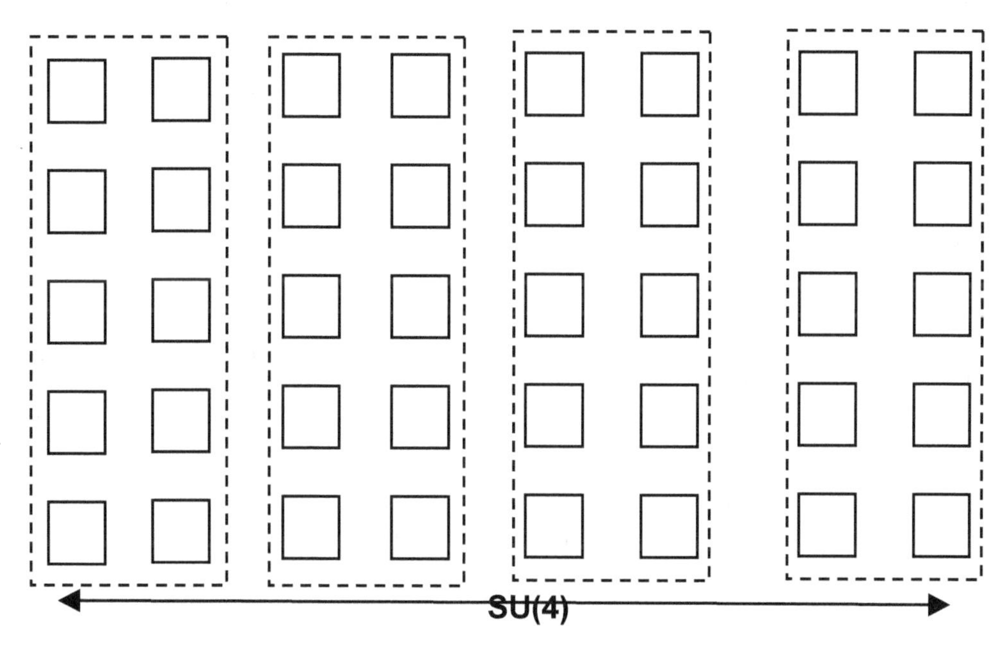

Figure 9-B.3. UTMOST has four QUeST copies. An SU(4) internal symmetry Connection group maps between corresponding fermions in the four copies: fermion by fermion. An additional U(1) Connection group applies to every corresponding fermion. It is not shown in this figure.

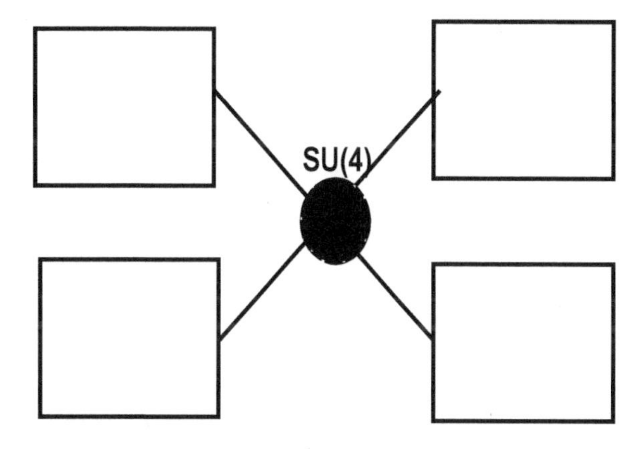

Figure 9-B.4. The SU(4) Connection Group of UTMOST connecting fermions in the four QUeST "copies" blocks. An additional U(1) Connection group applies to every corresponding fermion. It is not shown in this figure.

9-B.3 Maxiverse with Eight Real Coordinates (Dimensions)

The Maxiverse is a Blaha number N = 5 space with a dimension array of d_{d5} = 4096 dimensions.

The Maxiverse contains four copies of UTMOST. Each UTMOST has 4*6 = 24 space-time dimensions. (The remainder in each UTMOST copy consists of internal symmetries and Connection groups.) We allocate the 24 dimensions to eight real space-time dimensions plus 16 dimensions for a new ultraweak (possibly broken) SU(8) Connection group for the four parts of the Maxiverse. See Fig. 9-B.5.

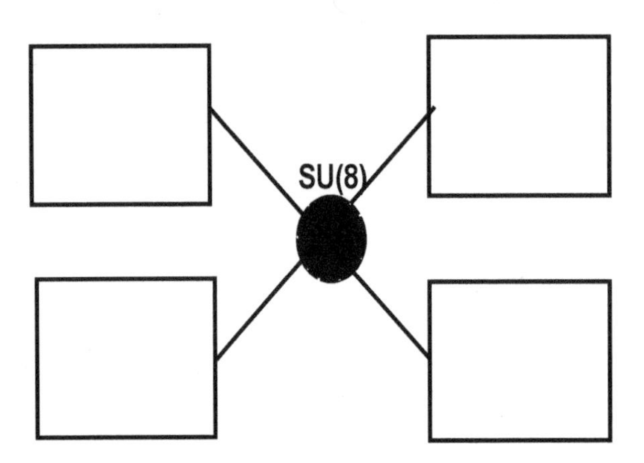

Figure 9-B.5. The SU(8) Connection Group of Maxiverse connecting fermions in the four UTMOST "copies" blocks.

9-B.4 Determining the Connection Groups for a Space

In the previous sections we determined the Connection Groups for N = 7, N = 6, and N = 5 spaces. The dimensions required for the Connection Groups of a space d_{gN} for $N \leq 6$ can be obtained by noting the space-time dimension of a space-time r is related to the space-time dimension r – 2 of the next lower space. Since a space is four copies of the next lower space it has 4(r – 2) space-time dimensions from the four duplicates. Then, using eq. 2.17, we find

$$4r_{N+1} = 4(r_N - 2) = r_N + d_{gN}$$

or

$$d_{gN} = 3r_N - 8 \qquad (9\text{-B.6})$$

where d_{gN} is the *real-valued* space-time dimensions reallocated to Connection Groups in space N. For N = 6, $r_6 = 6$, $d_{gN} = 10$. For N = 5, $r_5 = 8$, $d_{gN} = 16$. And so on.

Eq. 9-B.6 determines the dimensions available for Connection Groups for the space with Blaha number $N \leq 6$. Using eq. 9-B.1:

$$r_N = \log_2 (d_{dN}/16) \qquad (9\text{-B.7})$$

we relate d_{gN} to the size of the dimension array:

$$d_{gN} = 3 \log_2(d_{dN}/16) - 8 \tag{9-B.8}$$

thus specifying the number of space-time dimensions allocated to space number N.

10. UST Interactions

The UST (and HUST) symmetry groups have irreducible representations and interaction terms. We consider these features in this chapter. (Appendix C shows Complex General Relativity leads to a further U(4) group called the Species Group. This group generates interactions between all fermions and bosons. Thereby, it leads to the equality of inertial mass and gravitational mass – a known but previously unproven principle.)

10.1 Fermions in UST and HUST Symmetry Group Representations

We now consider the fermion composition of UST symmetry groups. The discussions are for free cases prior to the onset of interactions. Interactions lead to fermions acquiring parts of other fermions. Section 10.3 discusses UST and HUST interactions.

U(4) Generation Groups

The origin and form of the Generation Groups are described in Appendix 9-A and in Blaha (2015d). The four generations of fermions in each layer of each of the 8 fermion species have a four dimension irreducible representation. For example, e, μ, τ and an as yet unfound fourth generation lepton lie in a Generation Group irreducible representation. A similar representation exists for the neutrino species. Another example of a species set of fermions having a Generation Group representation is the quark species: q_{11}, q_{12}, q_{13}, and an as yet unfound fourth generation q_{14} quark. These quarks lie in another Generation Group irreducible representation. The Generation Group transformations "rotate" amongst these quarks (and also for leptons) creating superpositions of particle generations. Similarly for the other two quark species. Due to symmetry breaking effects each fermion acquires a part of the other fermions.

Each layer of fermions (four layers in the UST; eight layers in the r = 6 space-time universe HUST) has a different Generation Group with eight similar fermion species representations.

The Dark sector (if there is one) has a similar set of Generation Groups and species representations.

There are eight UST Generation Groups: four Normal sector Generation Groups and four Dark sector Generation Groups (if there is a Dark sector).

All fermions of a species in each layer have identical quantum numbers except for a Generation number that may be taken to have the values 1, 2, 3, 4 thus labeling the generation of each fermion within the representation. The Generation Group number is a sum (with coefficients) of diagonal components of the four Generation Group generator diagonal matrices.

Since we view the dimension arrays of higher space-time dimension spaces as having UST equivalents that we call HUSTs, and since dimension arrays quadruple in size as we go to higher dimensions in jumps of r of two dimensions, the above

Generation Group discussion applies in HUSTs. Each HUST has some multiple of fermion layers. For example, the r = 6 dimension array has 8 layers of Normal and Dark fermions. The 8 layers are separated into a pair of four layers where each is part of a duplicate UST structure. Thus the r = 6 case HUST becomes four UST copies (supplemented by a Connection Group. (See section 9-B.2.)

Figure 10.1 shows the UST Generation representations.

Layer Groups

The origin and form of the Layer Groups are described in Appendix 9-A and in our earlier books such as Blaha (2016a). A Layer group representation's fermions straddle layers. Thus its name. Generation by generation separately, each of the four generations of fermions in the four layers (for each of the 8 fermion species) forms a four dimension irreducible representation. See Fig. 10.2. For example, the four electron e-like fermions of the four layers lie in a Layer Group irreducible representation. And the four muon μ-like fermions of the four layers lie in a different Layer Group irreducible representation. Thus there are four Layer Group representations of charged leptons – one Layer Group for each generation. Similarly there are four Layer Group representations of neutral leptons. The six quark species each have four irreducible Layer Group representations. There are four Normal sector and four Dark sector Layer groups – one group for each generation of each sector.

The Layer Group transformations "rotate" amongst the representations of each species separately for each generation.

The Dark sector (if there is one) has a similar set of Layer Groups and species representations.

There are eight UST Layer groups: four Normal sector Layer Groups and four Dark sector Layer Groups (if there is a Dark sector). We place them for convenience in separate layers as in Fig. 9.1.

Due to Layer Group symmetry breaking effects each fermion of each Layer group representation acquires parts of the other fermions in the representation.

All fermions in a Layer Group representation have identical quantum numbers including the Generation number) except for a Layer number that may be taken to have the values 1, 2, 3, and 4, which label the layer of each fermion in the representation. The Layer number is a sum (with coefficients) of the diagonal components of four Layer Group generator diagonal matriccs.

Since we view the dimension arrays of Higher space-time dimension spaces as having UST equivalents that we call HUSTs, and since dimension arrays quadruple in size as we go to higher dimensions in jumps of r of two dimensions, the above Layer Group discussion applies in HUSTs. Each HUST has some multiple of fermion layers. For example, the r = 6 dimension array has 8 layers of Normal and 8 layers of Dark fermions. We view the r = 6 HUST as composed of four separate USTs. Thus the Generation Groups and the Layer Groups of each UST is separate from those of the other three USTs. The Connection Groups span the USTs as discussed below.

Figure 10.2 indicates the UST Layer representations spread across layers.

Connection Groups

The Connection Groups (Appendix 9-B) connect the fermions of the various layers of the Normal and Dark sectors. They are allocated in the UST in support of the principle:

A fermion in any block has interactions either directly, or indirectly, with every other fermion in every other block. (10.1)

Fermions, that are not so connected, are disjoint from the UST fermion spectrum and thus independent of Physics since they cannot be created or destroyed in Physical interactions. Effectively, they don't exist. (However they may have a gravitational interaction and thus may be created in pairs from highly energetic graviton-graviton interactions. But if so created they would be independent of elementary particle interactions. The question of the nature of Dark matter is still open. Unconnected Dark Matter would violate the principle in eq. 10.1.)

In the UST there are seven U(2) Connection groups. See Fig. 9-B.2. They are generated from seven units of four dimensions as shown in Fig. 9-B.1. The four groups in Fig. 9-B.2 labeled A, B, C, D connect the four layers in the Normal and Dark sector. They connect pairs of fermions (with one fermion in the Normal sector and one fermion in the Dark sector) in two dimension U(2) irreducible representations with each having exactly the same quantum numbers with one exception: there is a Darkness quantum number[59] that can be chosen to have the value one for Normal fermions and 2 for Dark fermions with negative values for anti-fermions.[60] The Darkness number is conserved in interactions. .

The three groups in Fig. 9-B.2 labeled E, F, G in both the Normal and Dark sectors connect four layers *vertically* in the Normal and Dark sector. They connect pairs of fermions (in two dimension U(2) irreducible representations) with exactly the same quantum numbers with one exception: the Layer quantum number[61] described in the Layer Groups discussion above. One of a fermion pair is in one layer; the other is in the adjoining layer. The Layer number for antifermions can be chosen to have the opposite values to that of the corresponding fermions.

A HUST is composed of quadruples of the next lower (in r space-time dimension) HUST. For r = 6 (the UTMOST) the HUST dimension array consists of four copies of UST. Thus the Connection groups of the r = 6 HUST consist of copies of those in the UST *plus* another overall SU(4) Connection Group connecting fermions in the four UST copies. See Fig. 9-B.3 and 9-B.4. The SU(4) Connection Group has 4 dimension irreducible representations for each quadruple of fermions from the four UST copies. The quadruples have identical quantum numbers except that each fermion in a quadruple has a different number, the HUST6 number, labeling the UST copies with values 1, 2, 3 ,4.

---

[59] Described in earlier books by the author.
[60] A similar Darkness numbering would hold for bosons.
[61] Described in earlier books by the author.

The Maxiverse (r = 8) consists of four UTMOST copies *plus* an SU(8) or an SU(4)⊗SU(4) Connection Group. See Fig. 9-B.5. The SU(8) Connection Group has 8 dimension irreducible representations.

The SU(4) factors have 4 dimension irreducible representations. We define one SU(4) for the Normal sector of the Maxiverse dimension array and the other SU(4) for the Dark sector. Each SU(4) representation has one corresponding fermion from each UTMOST part for the Normal and Dark parts separately.

The pairs within each quadruple have identical quantum numbers except that each fermion in a quadruple has a different number, the HUST8 number, labeling the fermion with one of the values 1, 2, 3 , 4 for the UTMOST of which it is a part. See Fig. 10.5.

SU(3)⊗U(1)

Each SU(3)⊗U(1) Strong Interaction group has a representation consisting of a quark triplet plus a lepton. Each group is different from the other SU(3)⊗U(1) groups in other layers in the Normal and Dark sectors. Quarks and leptons in different layers do not interact via the Strong Interactions. See Fig. 10.3 for the UST case. HUST cases are similar due to the quadrupling effect in HUSTs of higher space-time dimensions.

SU(2)⊗U(1)

Each SU(2)⊗U(1) ElectroWeak group has a representation consisting of a pair of quarks or a pair of leptons. These groups are different in each layer in the Normal and Dark sectors. See Fig. 10.3. See Fig. 10.4 for the UST case. The HUST cases are similar.

10.2 UST Fermion Interactions

Fermion interaction terms have the general form:

$$\overline{\psi}_f \gamma^\mu A_\mu \psi_i \tag{10.2}$$

where ψ_i represents an input fermion, and ψ_f represents an output fermion. A_μ represents a Yang-Mills vector boson. The "in" and "out" fermions may or may not have the same quantum numbers. The impact on fermions depends of whether the vector boson is "charged" in some quantum number(s).

Figs. 10.1 – 10.4 depict the representations of the various Internal Symmetry Groups. Due to the buildup of HyperCosmos space dimension arrays in quadruples of the previous dimension array the figures of Figs. 10.1 – 10.4 hold within both the UST and the HUSPs of the higher space-time dimensions.

We discuss fermion interactions below for "bare" fermions prior to mixing by interactions. The fermions that remain to be found experimentally must have extremely large masses. The gauge fields that provide the interactions that have not been found as yet also must have extremely large masses and/or very small coupling constants. The UST Higgs particles have been discussed in chapter 4. The Generation, Layer and

Connection Groups Higgs Mechanisms may be expected to be analogous to the Higgs Mechanism in the ElectroWeak case.

10.2.1 Generation Group Interactions

Generation Group interaction terms for each pair of ψ's in eq. 10.2 transform an input wave function to the output wave function where the only change in internal symmetry numbers is a "possible" change in the Generation number. Generation numbers are changed for A's that have non-zero Generation "charge." Generation numbers are not changed for A's that have zero Generation "charge."

10.2.2 Layer Group Interactions

Layer Group interaction terms for each pair of ψ's transform an input wave function to the output wave function where the only change in internal symmetry numbers is a "possible" change in the Layer number. Layer numbers are changed for A's that have non-zero Layer "charge." Layer numbers are not changed for A's that have zero Layer "charge."

10.2.3 Connection Group Interactions

"Horizontal"[62] Connection Group interaction terms for each pair of ψ's transform an input wave function to the output wave function where the only change in internal symmetry numbers is a "possible" change in the Darkness number. Darkness numbers are changed for A's that have non-zero Darkness "charge." Darkness numbers are not changed for A's that have zero Darkness "charge."

"Vertical"[63] Connection Group interaction terms for each pair of ψ's transform an input wave function to the output wave function where the only change in internal symmetry numbers is a "possible" change in the Layer number. Layer numbers are changed for A's that have non-zero Layer "charge." Layer numbers are not changed for A's that have zero Layer "charge."

10.2.4 SU(3) Group Interactions

These interactions are the same as those in The Standard Model.

10.2.5 SU(3) Group Interactions

These interactions are the same as those in The Standard Model with the addition of a fourth generation in each of the four UST layers.

10.3 The 4 ×4 Cabibbo-Kobayashi-Maskawa Matrix

The form of the four generation CKM matrix appears in Fig. 10.6. The CKM matrix reflects the impact of interactions on the quark spectrum. We may view the CKM matrix as generated from the combined effect of the Generation Group and Layer Group interactions. The CKM matrix denoted C satisfies

$$|C|^2 = 1 \qquad (10.3)$$

---

[62] Groups A, B, C, D in Fig. 10.3.

[63] Groups E, F, G in Fig. 10.3.

The form of the matrix is

$$C = S_{Gendown}{}^{\dagger} S_{Laydown}{}^{\dagger} S_{Genup} S_{Layup} \tag{10.4}$$

where S_{Genup} and S_{Layup} are matrices from the Generation and Layer Groups respectively for an up-type quark U(4) representation and where $S_{Gendown}$ and $S_{Laydown}$ are matrices of the Generation and Layer Groups respectively for a down-type quark U(4) representation. The quarks of each of these quark species appear labeling rows and columns in Fig. 10.6.

The CKM matrix reflects contributions from the Generation Group and to a lesser extent from the Layer Group. The large masses of both groups gauge fields determine their relative impact. The impact of the Layer Group is muted due to its necessarily extremely large gauge vector boson masses and very small coupling constants. If this were not the case the higher layers of fermions might have been detected in experiments.

Thus we approximate C with

$$C \cong S_{Gendown}{}^{\dagger} S_{Genup} \tag{10.5}$$

Up to an overall phase since $S_{Layup} \approx S_{Laydown} \approx I$ the identity element. Thus we reduce C to a product of U(4) matrices which gives a SU(4) matrix (eq. 10.3) with an appropriate choice of phase.

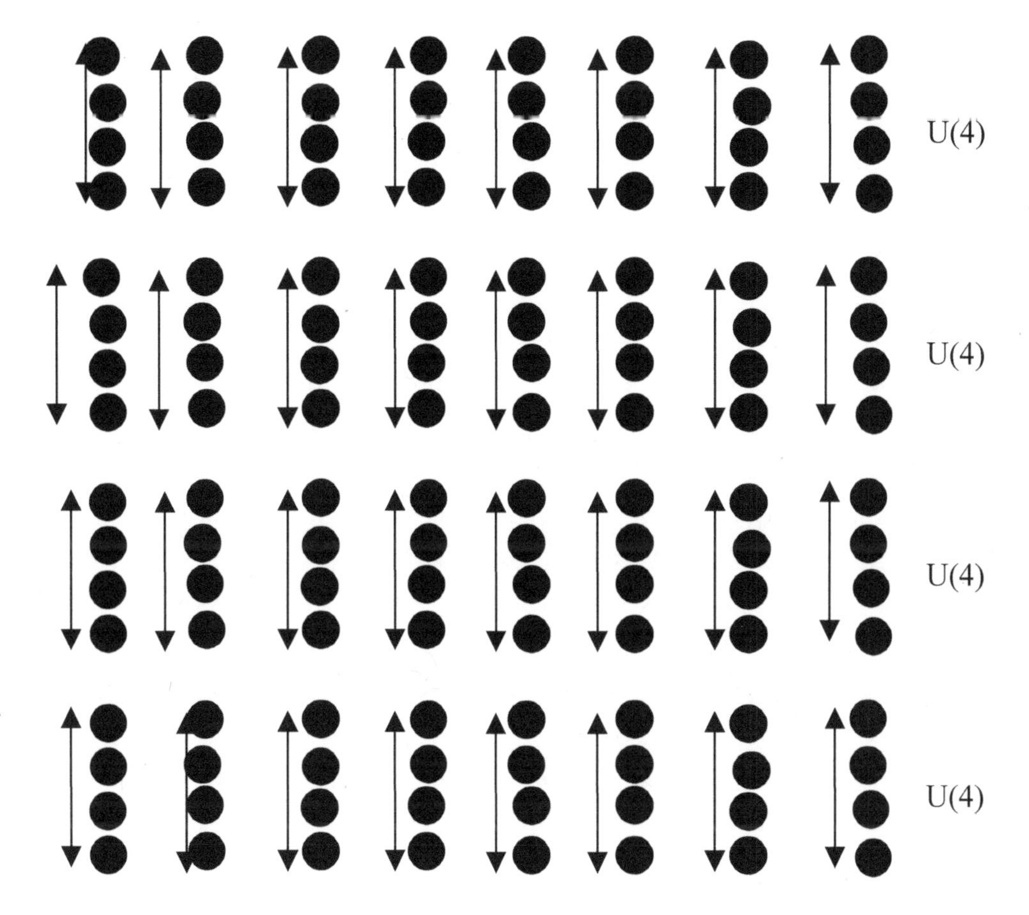

Figure 10.1. There is one U(4) Generation group for each layer in the Normal sector and also in the Dark sector. Each species column in each layer is a U(4) irreducible representation. There are four Generation groups in the Normal sector and four Generation groups in the Dark sector totalling to 8 UST Generation Groups. representation. Each of the 8 species in each layer furnishes a separate Generation Group representation.

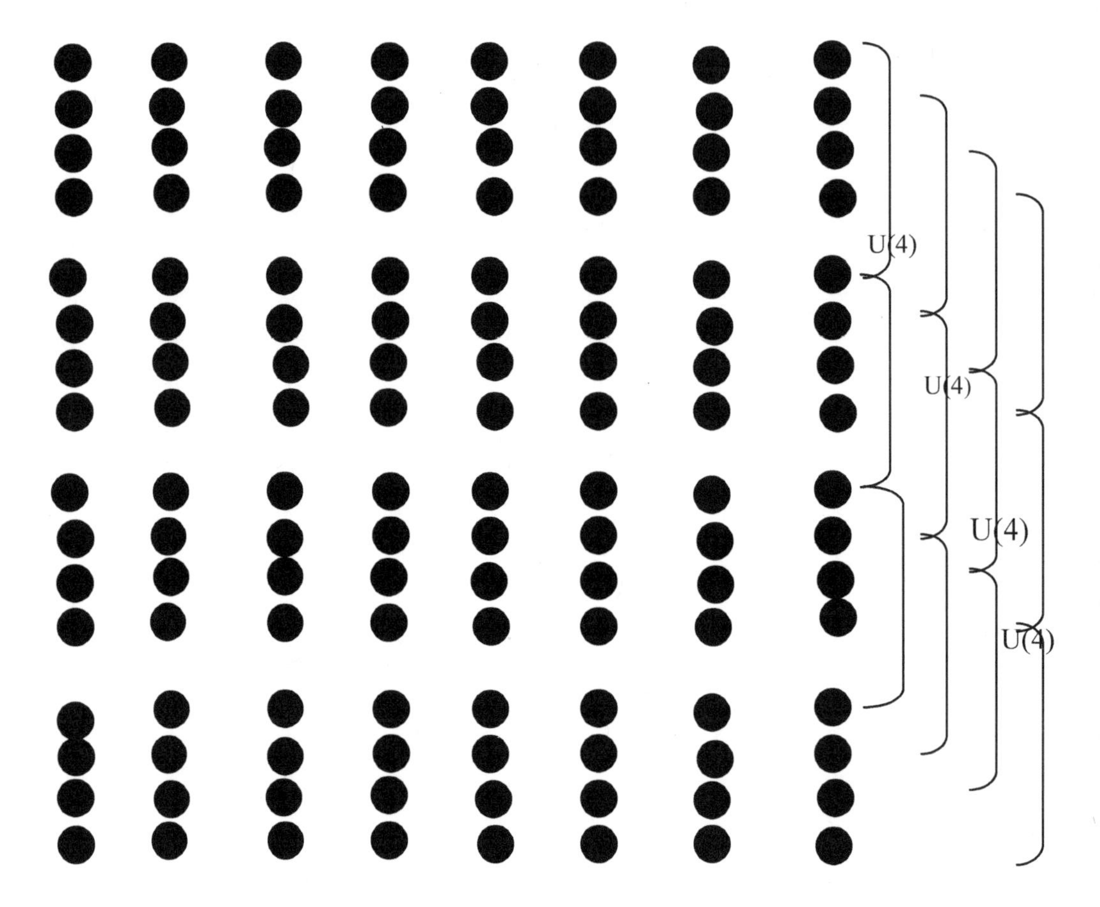

Figure 10.2. Four U(4) Layer groups – one for each generation. For symmetry we place each Layer group in a different layer. There are four Layer groups in both the Normal and Dark sector totalling to 8 Layer groups. Each vertical line of symbols on the right indicates the four fermions in a Layer representation. Each of the 8 species furnishes a separate Layer Group representation for each generation totalling four representations per species.

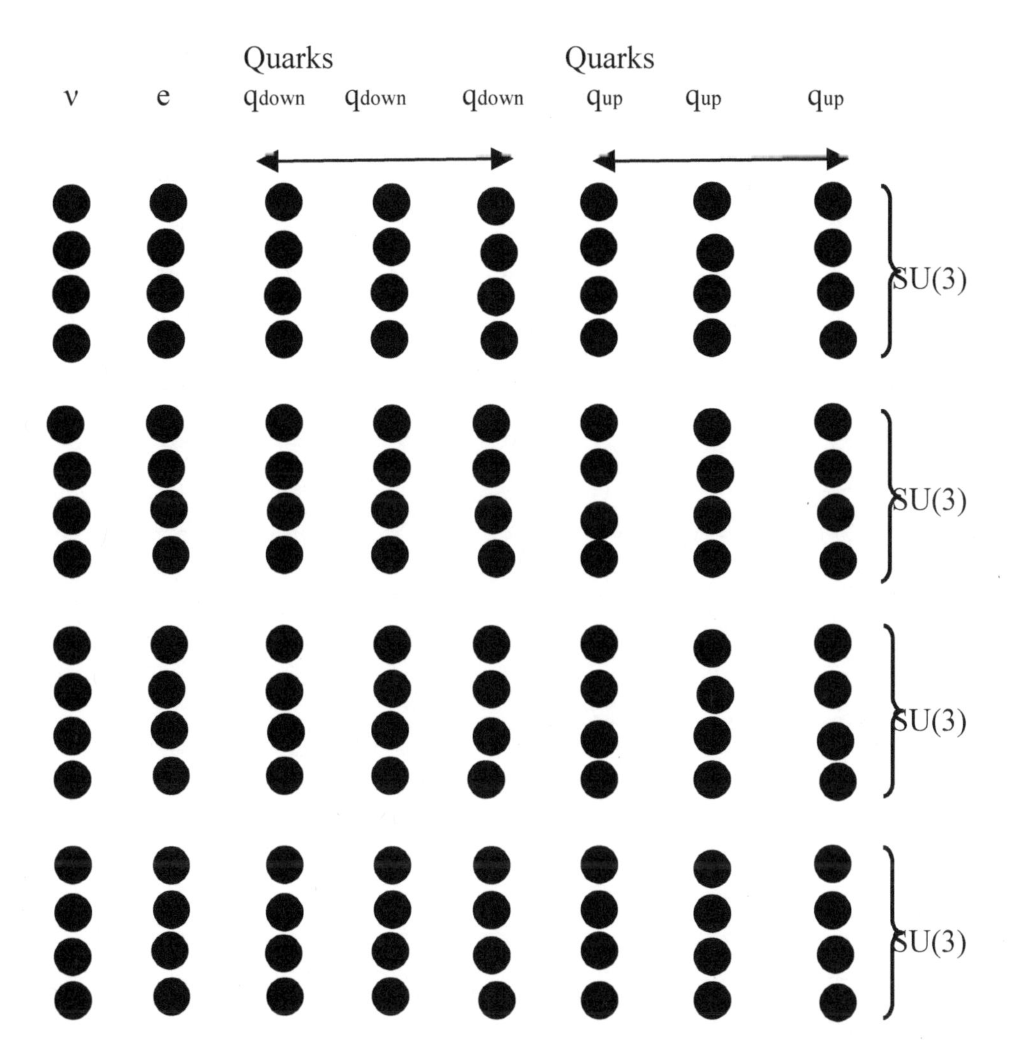

Figure 10.3. Four SU(3) Strong interaction groups in the Normal sector and four SU(3) groups in the Dark sector. Interactions are between any quark of any generation within each layer in the Normal sector and also in the Dark sector.. There is a different SU(3) for each layer.in the Normal and Dark sectors totally to 8 SU(3)'s.

Figure 10.4. Four SU(2)⊗U(1) Weak Interaction groups in the Normal sector and four SU(2)⊗U(1) Weak Interaction groups in the Dark sector. There is a different SU(2)⊗U(1) for each layer in the Normal sector and also in the Dark sector.

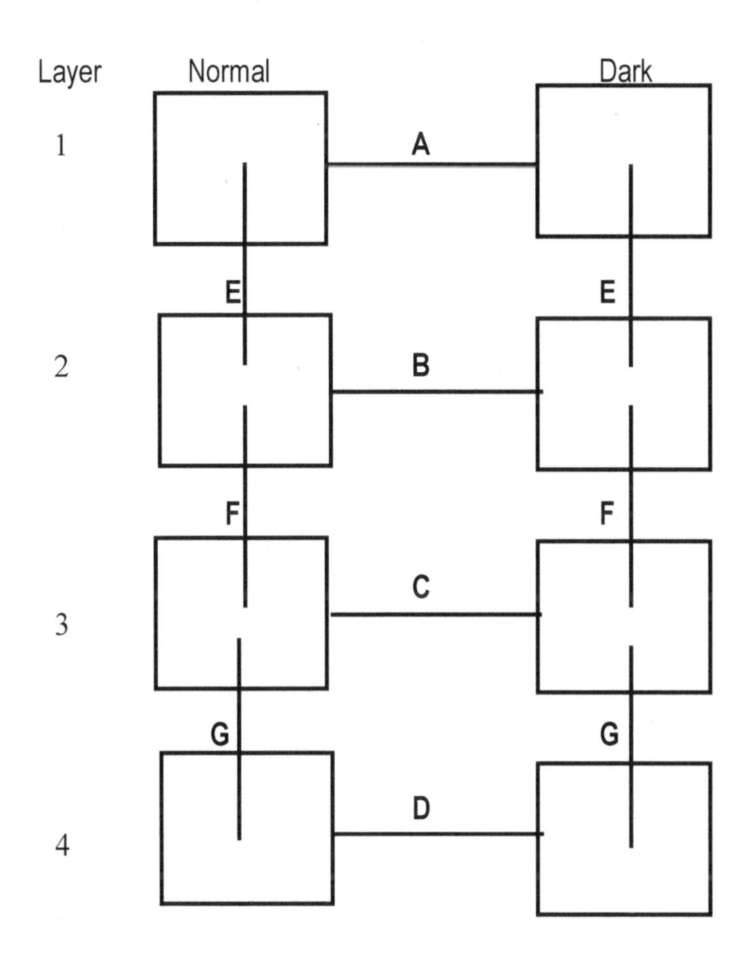

Figure 10.5. The three vertical U(2) Connection groups[64] (shown as 3 lines: E, F, G) between the QUeST/UST blocks in the N = 7 HyperCosmos; and the four horizontal U(2) Connection groups (shown as 4 lines: A, B, C, D) between the QUeST/UST blocks The Connection groups generate rotations and interactions between corresponding fermions of each pair of blocks. There are 7 UST Connection groups.

[64] Connection groups are discussed in Appendix 9-B.

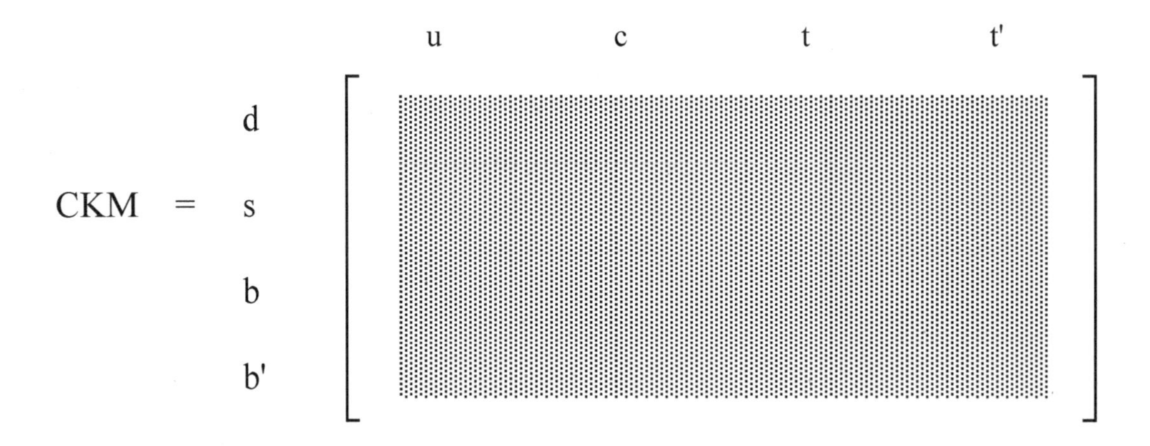

Figure 10.6. The 4×4 UST Cabibbo-Kobayashi-Maskawa Matrix for quarks. Quark symbols corresponding to the rows and columns are displayed. The quarks b' and t' remain to be found.

11. Universe Generation from Primordial Universes

11.1 Universe Generation from Primordial Universes

Universes may be generated from the wave functions of the Geometric ProtoCosmos presented in Blaha (2023c). The Geometric ProtoCosmos Model (GPM) constructed the four energy eigenfunctions per energy level geometrically from the spins of PseudoFermions.

In this chapter we will construct the eigenfunction for each energy level using PseudoQuantum PseudoFermion wave functions. Each energy level will have two PseudoFermion wave functions. The wave functions will embody two spaces: a parent space of space-time dimension r and a child space of space-time dimension r'. Each space will have an independent dimension array d_{dN} that supports the space-time and internal symmetries of the space.

Unlike the Geometric ProtoCosmos Model of Blaha (2023c) we define a new model that will give each spin creation/annihilation operator a Cayley number index that will lead to the independent dimension arrays of the parent space and child space. This model, which we call the Coupled Universes Theory (CUT), is only partly geometric. Its advantage over GPM is based on the independence of parent and child dimension arrays. In the GPM there was only one joint dimension array for parent and child.

While there are a number of possible CUT scenarios we will use a scenario where the lowest "energy" ProtoCosmos bound state solution's universe part (having the largest space-time dimensions and the largest dimension array) is the source of all subsequent universes. We begin by defining the GPM. Then we define a new form for PseudoFermions – Independent PseudoFermions (IPF) and then derive the CUT Lagrangian Model.[65] Our universe, and its Hubble expansion, is discussed as well as hierarchies and networks of universes.

11.1.1 PseudoFermions[66]

Following Blaha (2023c) we define[67] PseudoFermions with the form

$$\psi_{i\alpha\beta}(y, z) \tag{11.1}$$

where y and z are independent coordinates in r space-time dimensions, where i = 1, 2 labels the PseudoQuantum fields, and α and β are spinor indices.[68]

---

[65] We choose the HyperCosmos space version initially with the expectation that the Dark sector, which it has, will eventually be found.

[66] From Blaha (2023c).

[67] We follow the conventions of Bjorken (1965) with the $g^{\mu\nu}$ metric (1, -1, -1, -1).

[68] Appendix C of Blaha (2016f) presents a first quantized PseudoQuantum theory CQ Mechanics that embodies both classical and quantum theory. **This theory is the non-relativistic quantum mechanics limit of the PseudoFermion theory developed here.** CQ Mechanics has two sets of coordinates that combine to create a generalization of

We begin by defining a free PseudoFermion PseudoQuantum Lagrangian with two related wave functions ψ_1 and ψ_2 that are functions of two sets of coordinates in r space-time dimensions, y and z,

$$\mathscr{L} = \overline{\psi}_{2\alpha\beta}[-M^{-1}\gamma_{y\alpha\kappa}{}^{\mu}\cdot\partial/\partial y^{\mu}\ \gamma_{z\beta\lambda}{}^{\nu}\partial/\partial z^{\nu}\ - M]\psi_{1\kappa\lambda} + \\ + \overline{\psi}_{1\alpha\beta}[-M^{-1}\gamma_{y\alpha\kappa}{}^{\mu}\cdot\partial/\partial y^{\mu}\ \gamma_{z\beta\lambda}{}^{\nu}\partial/\partial z^{\nu} - M]\psi_{2\kappa\lambda} \tag{11.2}$$

where $\gamma_y{}^{\mu}$ and $\gamma_z{}^{\mu}$ are Dirac matrices for y and z coordinates respectively, y and z are coordinates in r and r' dimension space-times respectively, M is the mass, and

$$\overline{\psi}_{i\alpha\beta} = \psi_{i\kappa\lambda}{}^{\dagger}\ \gamma_y{}^0{}_{\kappa\alpha}\ \gamma_z{}^0{}_{\lambda\beta} \tag{11.2a}$$

for i = 1, 2 where the subscripts y and y indicate Dirac spinors associated with the y and z coordinates.

The equations of motion are

$$[-M^{-1}\gamma_y{}^{\mu}\cdot\partial/\partial y^{\mu}\ \gamma_z{}^{\nu}\partial/\partial z^{\nu}\ - M]\psi_1 = 0 \tag{11.2a}$$
$$[-M^{-1}\gamma_y{}^{\mu}\cdot\partial/\partial y^{\mu}\ \gamma_z{}^{\nu}\partial/\partial z^{\nu}\ - M]\psi_2 = 0 \tag{11.2b}$$

We define subsidiary equations of motion

$$[i\gamma_y{}^{\mu}\cdot\partial/\partial y^{\mu}\ - M]\psi_j = 0 \tag{11.2c}$$
$$[i\gamma_z{}^{\nu}\partial/\partial z^{\nu}\ - M]\psi_j = 0 \tag{11.2d}$$

for j = 1, 2. Eqs. 11.2c and 11.2d imply eqs. 11.2a and 11.2b.

One conjugate momentum is (with two indices α, β)

$$\pi_{y1\alpha\beta} = \partial\mathscr{L}/\partial(\partial\psi_{1\alpha\beta}/\partial y^0) = -M^{-1}(\gamma_z{}^{\nu}\partial/\partial z^{\nu}\,\gamma_z{}^0\psi_2{}^{\dagger})_{\alpha\beta} = (\psi_2{}^{\dagger}\gamma_z{}^0)_{\alpha\beta} \tag{11.3}$$

after partial integrations (with surface terms having the value zero) of

$$L = \int d^r y \int d^r z\ \mathscr{L}$$

using the subsidiary equation of motion:

$$\gamma_z{}^{\nu}\partial/\partial z^{\nu}\ \psi_2{}^{\dagger} = -M\psi_2{}^{\dagger} \tag{11.4}$$

---

conventional Quantum Mechanics. It has appications in a generalized Feynman path integral formalism, a generalized Schrödinger equation, a generalized Boltzmann equation, the Fokker-Planck equation, a generalized approach to quantum and classical chaos, and to quantum entanglement as well as semi-quantum entanglement. Our "Pseudo" formalisms apply to both Quantum Field Theory and Quantum Mechanics. In these applications there is a clear almost continuous transition between the quantum to the classical sectors.

Similarly the other conjugate momenta are

$$\pi_{z1\alpha\beta} = \partial\mathscr{L}/\partial(\partial\psi_{1\alpha\beta}/\partial z^0) = -M^{-1}(\gamma_y{}^v\partial/\partial y^v\gamma_y{}^0\psi_2{}^\dagger)_{\alpha\beta} = (\psi_2{}^\dagger\gamma_y{}^0)_{\alpha\beta} \qquad (11.5)$$

and

$$\pi_{y2\alpha\beta} = \partial\mathscr{L}/\partial(\partial\psi_{2\alpha\beta}/\partial y^0) = -M^{-1}(\gamma_z{}^v\partial/\partial z^v\psi_1{}^\dagger)_{\alpha\beta} = (\psi_1{}^\dagger\gamma_z{}^0)_{\alpha\beta} \qquad (11.6)$$

$$\pi_{z2\alpha\beta} = \partial\mathscr{L}/\partial(\partial\psi_{2\alpha\beta}/\partial z^0) = -M^{-1}(\gamma_z{}^v\partial/\partial y^v\psi_1{}^\dagger)_{\alpha\beta} = (\psi_1{}^\dagger\gamma_y{}^0)_{\alpha\beta} \qquad (11.7)$$

using the subsidiary equations of motion:

$$\gamma_z{}^v\partial/\partial z^v\psi_1{}^\dagger = -M\psi_1{}^\dagger \qquad (11.8)$$

$$\gamma_y{}^v\partial/\partial y^v\psi_1{}^\dagger = -M\psi_1{}^\dagger \qquad (11.9)$$

$$\gamma_y{}^v\partial/\partial y^v\psi_1{}^\dagger = -M\psi_1{}^\dagger \qquad (11.10)$$

We define new momenta to preserve the y – z symmetry of section 11.1:

$$\pi_{1\alpha\beta} = (\pi_{y1}\gamma_z{}^0)_{\alpha\beta} = (\pi_{z1}\gamma_y{}^0)_{\alpha\beta} = \psi_2{}^\dagger{}_{\alpha\beta} \qquad (11.11)$$

$$\pi_{2\alpha\beta} = (\pi_{y2}\gamma_z{}^0)_{\alpha\beta} = (\pi_{z2}\gamma_y{}^0)_{\alpha\beta} = \psi_1{}^\dagger{}_{\alpha\beta} \qquad (11.12)$$

The form of the conjugate momenta implies the only non-zero equal time anticommutators are:[69]

$$\{\pi_{j\alpha\beta}(\mathbf{y}, y^0, \mathbf{z}, z^0),\ \psi_{i\kappa\lambda}(\mathbf{y}', y^0, \mathbf{z}', z^0)\} = \{\psi_{j\alpha\beta}{}^\dagger(\mathbf{y}, y^0, \mathbf{z}, z^0),\ \psi_{i\kappa\lambda}(\mathbf{y}', y^0, \mathbf{z}', z^0)\}$$

$$= (1-\delta_{ij})\,\delta_{\alpha\kappa}\,{}_\beta\delta_{\beta\lambda}\,\delta^{r-1}(\mathbf{y}-\mathbf{y}')\delta^{r'-1}(z-\mathbf{z}') \qquad (11.13)$$

for i, j = 1, 2.

The free PseudoFermion wave function has the form:

$$\psi_{i\alpha\beta}(y, z) = \sum_{s_1}\sum_{s_2}\int dp^{r-1}\int dq^{r'-1}\,N(p, q)\,[b_i(p, q, s_1, s_2)u_\alpha(p,s_1)u_\beta(q, s_2)\exp(-ip\cdot y - iq\cdot z) +$$

$$+ d_i{}^\dagger(p, q, s1, s2)v_\alpha(p,s1)v_\beta(q, s2)\exp(ip\cdot y + iq\cdot z)] \qquad (11.14)$$

plus Hermitean conjugates for i = 1, 2 where N(p, q) is a normalization factor.

The creation and annihilation operators satisfy the anticommutation relations:

$$\{b_i(p, q, s_1, s_2), b_j{}^\dagger(p', q', s_1', s_2')\} = (1-\delta_{ij})\,\delta_{s_1,s_1'}\delta_{s_2,s_2'}\delta^{r-1}(\mathbf{p}-\mathbf{p}')\delta^{r'-1}(\mathbf{q}-\mathbf{q}') \qquad (11.15)$$

$$\{d_i(p, q, s_1, s_2), d_j(p', q', s_1', s_2')\} = (1-\delta_{ij})\,\delta_{s_1,s_1'}\delta_{s_2,s_2'}\delta^{r-1}(\mathbf{p}-\mathbf{p}')\delta^{r'-1}(\mathbf{q}-\mathbf{q}') \qquad (11.16)$$

The other anticommutators have zero values.

They lead to the equal time anticommutation relations of eq. 11.13. The result is the free PseudoFermion formalism.

---

[69] See S. Blaha, Il Nuovo Cimento **49A**, 35 (1979) for one coordinate system, PseudoQuantum fermions.

11.1.2 GPM ProtoCosmos Lagragian Model leading to HyperCosmos spaces[70]

We found the GPM[71] ProtoCosmos Lagrangian Model leading to HyperCosmos and Second Kind HyperCosmos spaces for four PseudoQuantum fields for each type is

$$\begin{aligned}\mathscr{L} = \overline{\psi}[\gamma_x{}^{\mu}\, \partial/\partial x^{\mu} + m + V(x) \; - \\ - M^{-1}\gamma_y{}^{\mu}\partial/\partial y^{\mu}|_{f1(E)}\, \gamma_z{}^{\nu}\partial/\partial z^{\nu}|_{f1'(E)} + 2M \; - \\ - M^{-1}\gamma_y{}^{\mu}\partial/\partial u^{\mu}|_{f1(E)}\, \gamma_z{}^{\nu}\partial/\partial v^{\nu}|_{f2(E)}]\psi \; + \\ + \sum_{i=1}^{4} {}^1\overline{\psi}_{i\alpha\beta}(y, z)[-M^{-1}\gamma_{y\alpha\kappa}{}^{\mu}\partial/\partial y^{\mu}|_{f1(E)}\, \gamma_{z\beta\lambda}{}^{\nu}\partial/\partial z^{\nu}|_{f1'(E)} + M]\, {}^1\psi_{i\kappa\lambda}(y, z) + \\ + \sum_{i=1}^{4} {}^2\overline{\psi}_{i\alpha\beta}(u, v)[-M^{-1}\gamma_{y\alpha\kappa}{}^{\mu}\partial/\partial u^{\mu}|_{f1(E)}\, \gamma_{z\beta\lambda}{}^{\nu}\partial/\partial v^{\nu}|_{f2(E)} + M]\, {}^2\psi_{i\kappa\lambda}(u, v) + \text{h.c.}\end{aligned} \tag{11.17}$$

where x is a one dimension coordinate, $\gamma_x{}^{\mu}$ is the Dirac matrix for the x coordinates, m is the mass in the x coordinate space, and M is the mass in the four spaces labeled with coordinates: y, z, u, and v. Each of the four coordinate spaces has a set of Dirac matrices labeled correspondingly with subscripts. In accordance with the PseudoFermion discussions of sections 4.5 and 4.6 of Blaha (2023c) there are four wave functions ${}^1\psi_{...}$ that lead to the HyperCosmos spectrum of spaces, and four wave functions ${}^2\psi_{...}$ that lead to the Second Kind HyperCosmos spectrum of spaces.

When the dynamic equations are generated from the Lagrangian the x coordinate equation will generate a geometric energy spectrum E. Each energy determines the space-time dimension of the corresponding HyperCosmos space and Second Kind HyperCosmos space and thereby the $\gamma \cdot \partial$ factors in the dynamic equations. The functions specifying the space-time dimensions are

$$f1(E) = r = \ln_2[(m - E)/16] \tag{11.18}$$

In Blaha (2023c) we considered the possibility that r' would be equal to r + 4 with the result that the field $\psi_{i\alpha\beta}(y, z)$ would be a HyperUnification space fermion field. We set

$$f1'(E) = r' = 4 + f1(E) = \ln_2(m - E)$$
$$f2(E) = r' = 4 + \ln_2[(m - E)/8] = \ln_2[2(m - E)]$$

They were determined by requiring the PseudoFermion solutions of the dynamic equations to give the dimension arrays as seen in sections 4.5 and 4.6 of Blaha (2023c).

[70] From Blaha (2023c).
[71] Blaha (2023c)

11.1.3 Independent PseudoFermions (IPF) and Coupled Universes Theory (CUT)

GPM provides a set of solutions that each contain PseudoFermion wave functions for universes. These PseudoFermion wave functions define a composite dimension array for the entire space-time dimension of the PseudoFermion.

We now develop a new type of PseudoQuantum PseudoFermion – the Independent PseudoFermion (IPF) – that implements the Coupled Universes Theory (CUT) giving separate dimension arrays to the parent universe and to its child universe(s). It provides a better alternative to GPM. It begins with the new CUT ProtoCosmos Lagrangian:

$$\begin{aligned}\mathscr{L} = \overline{\psi}[\gamma_x{}^{\mu}\,\partial/\partial x^{\mu} + m + V(x) &- \\ - M^{-1}\gamma_y{}^{\mu}\partial/\partial y^{\mu}|_{f1(E)}\,\gamma_z{}^{\nu}\partial/\partial z^{\nu}|_{f1'(E)} &+ 2M - \\ - M^{-1}\gamma_y{}^{\mu}\partial/\partial u^{\mu}|_{f1(E)}\,\gamma_z{}^{\nu}\partial/\partial v^{\nu}|_{f2(E)}]\psi &+ \\ + \sum_{i=1}^{2} {}^{1}\overline{\psi}_{i\alpha\beta}(y, z)[-M^{-1}\gamma_{y\alpha\kappa}{}^{\mu}\partial/\partial y^{\mu}|_{f1(E)}\,\gamma_{z\beta\lambda}{}^{\nu}\partial/\partial z^{\nu}|_{f1'(E)} + M]\,{}^{1}\psi_{i\kappa\lambda}(y, z) &+ \\ + \sum_{i=1}^{2} {}^{2}\overline{\psi}_{i\alpha\beta}(u, v)[-M^{-1}\gamma_{y\alpha\kappa}{}^{\mu}\partial/\partial u^{\mu}|_{f1(E)}\,\gamma_{z\beta\lambda}{}^{\nu}\partial/\partial v^{\nu}|_{f2(E)} + M]\,{}^{2}\psi_{i\kappa\lambda}(u, v) &+ \text{h.c.}\end{aligned} \tag{11.19}$$

where the free IPF wave function has the form:

$$\begin{aligned}\psi_{i\alpha\beta}(y, z) = \sum_{s_1}\sum_{s_2}\int dp^{r-1}\int dq^{r'-1}\, N(p, q)\, [b_i(p, q, s_1, s_2)u_\alpha(p,s_1)u_\beta(q, s_2)\exp(-ip\cdot y - iq\cdot z) &+ \\ + d_i^{\dagger}(p, q, s1, s2)v_\alpha(p,s1)v_\beta(q, s2)\exp(ip\cdot y + iq\cdot z)]\end{aligned} \tag{11.21}$$

plus Hermitean conjugates for $i = 1, 2$ where $N(p, q)$ is a normalization factor.

The creation and annihilation operators satisfy the anticommutation relations of eqs. 11.15 and 11.16. The forms of the parent and child dimension arrays are described in section 11.1.3.1.

We will consider the 10 HyperCosmos spaces[72] that emerge from the ProtoCosmos defined by eq. 11.19. Physically we will assume that one 18 space-time dimension solution state "starts" the cascade of the HyperCosmos spaces. The $r = 18$ space-time dimension (universe) state supports a child space-time dimension r' universe. The Physically acceptable child r' space-time dimension values[73] are:

$$4, 6, 8, 10, 12, 14, 16 \tag{11.21a}$$

Any of these values specify a Physically acceptable child space that exists within the $r = 18$ space. The values $r' = 0, 2$ are also acceptable. But we rule them out since the

[72] Second Kind HyperCosmos spaces are also present.
[73] See Fig. 11.3.

universes that they imply do not seem to exist within our universe.[74] The degree of freedom allowed in the choice of r' is analogous to the choice of angular momentum L^2 values in the case of the Quantum Mechanics Hydrogen atom. The value of the principal quantum number n allows and limits L^2 values for each possible energy level. In the present case the Energy E, which determines the value of r, constrains possible values of the subspace space-time dimensions r' on Physical grounds.

We therefore set

$$f1'(E) = f2(E) = r' \tag{11.21b}$$

in the CUT ProtoCosmos spectrum derivation.

The dynamic equations obtained by variations with respect to fields are

$$[\gamma_x{}^{\mu}\, \partial/\partial x^{\mu} + m + V(x) - M^{-1}\gamma_y{}^{\mu}\partial/\partial y^{\mu}|_{f1(E)}\, \gamma_z{}^{\nu}\partial/\partial z^{\nu}|_{f1'(E)} + 2M - \\ - M^{-1}\gamma_y{}^{\mu}\partial/\partial u^{\mu}|_{f1(E)}\, \gamma_z{}^{\nu}\partial/\partial v^{\nu}|_{f2(E)}]\psi = 0 \tag{11.22}$$

$$[M^{-1}\gamma_y{}^{\mu}\partial/\partial y^{\mu}|_{f1(E)}\, \gamma_z{}^{\nu}\partial/\partial z^{\nu}|_{f1'(E)} - M]\, {}^1\psi_j(y, z) = 0 \tag{11.23}$$

$$[M^{-1}\gamma_y{}^{\mu}\partial/\partial u^{\mu}|_{f1(E)}\, \gamma_z{}^{\nu}\partial/\partial v^{\nu}|_{f2(E)} - M]\, {}^2\psi_j(u, v) = 0 \tag{11.24}$$

If we separate eq. 11.22:

$$\psi = \psi_0(x)\, {}^1\psi(y, z)\, {}^2\psi\, (u, v) \tag{11.25}$$

then eq. 11.22 – 11.25 imply the energy E eigenvalue equation:

$$[\gamma_x{}^{\mu}\, \partial/\partial x^{\mu} + m + V(x)]\psi_0 = H\psi_0(y,z,x) = E\psi_0(y, z, x) \tag{11.26}$$

The solutions have the form:

$$\psi = \psi_0\, [a\, {}^1\psi_1(y, z) + b\, {}^1\psi_2(y, z)][a_2\, {}^2\psi_1(u, v) + b_2\, {}^2\psi_2(u, v)] \tag{11.27}$$

where a, b, a_2, b_2 are constants.

The HyperCosmos spectrums appear in Figs. 11.1 and 11.2. In general the CUT spaces have the spectrum of r' values listed in Fig. 11.3.

11.1.3.1 IPF CUT HyperCosmos Introduction of Internal Symmetries

In section 4.1 of Blaha (2003c) we generated the joint dimension array from the combined effect of the y and z spaces. In this section we require the parent space and the child space to each have their own independent dimension arrays. The respective parent and child spins both generate spin factors. For a pair of PseudoQuantum wave functions we find the number of spin states is[75]

---

[74] The anyon in section 5.2.2 is a possible exception.

[75] The PseudoFermion of eq. 11.14 above, which is restricted to two PseudoQuantum wave functions, has $2^{r/2 - 1}$ spin values for the spins labeled s_1. Similarly it has $2^{r'/2 - 1}$ spin values for the spins labeled s_2. For each spin value there

$$N_s = 2^{r/2 + r'/2} \tag{11.28}$$

Our goal is: two independent dimension arrays: one for the parent space/universe with the dimension array size $d_d = 2^{r+4}$ and one for the child space/universe $d_d = 2^{r'+4}$.

We now make each fermion creation/annihilation operator a Cayley number[76] operator of Cayley number $2^{r/2+ r'/2+ 8}$. This decision is an assumption designed to obtain the HyperCosmos form of the size of dimension arrays.[77] We find

$$\begin{aligned} N_{stot} &= 2^{r/2 + r'/2 + 8} N_s \\ &= 2^{r + 4 + r' + 4} \\ &= d_{dr} d_{dr'} \end{aligned} \tag{11.29}$$

where d_{dr} is the dimension array size for space-time dimension r. Eq. 11.29 gives the desired dimension array sizes for the CUT parent and child spaces and universes. *It implements the required form of the PseudoQuantum IPF wave functions.* The result is the CUT PseudoQuantum ProtoCosmos theory with two wave functions and the Lagrangian of eq. 11.19.

11.1.3.1.1 Space-Times and Dimension Arrays

The child space space-time r' of an IPF wave function describes the IPF dynamics in the child universe within the parent universe. The child space dimension array contains a set of dimensions that includes dimensions for an internal space-time of a universe. We set that internal universe space-time dimension to r'. Particles in the child universe have wave functions using the r' internal universe space-time dimensions.

Thus there are three sets of space-time dimensions in this case: parent space-time r dimensions, child space-time r' dimensions, and child space universe internal space-time r' dimensions. The latter two sets of space-time dimensions are independent although we have assigned the same number of space-time dimensions to them.

11.1.3.1.2 Universe Generation

We now view the HyperCosmos as being generated from the r = 18 CUT energy solution's IPF child states. An IPF state as shown above has one child universe. (It is possible to define IPF's with any number of child universes.) Subsequently these states cause the generation of subspaces and their universes of the space-time dimensions in eq. 11.20a. Thus the r = 18 space and universe generates a cascade of subspaces and subuniverses.

---

are 2^2 b's and d's (namely b_1, $b_1^\dagger$, b_2, $b_2^\dagger$, d_1, $d_1^\dagger$, d_2, $d_2^\dagger$) taking account of the two PseudoQuantum fields for each HyperCosmos and 2nd Kind HyperCosmos case. Thus the total number of b's and d's is given by 11.28.

[76] The Cayley numbers have the values $C_n = 2^n$.

[77] Both the parent and the child creation/annihilation operators must have the same Cayley number factor because of the form of the operators, which embody both parent and child, in eq. 11.21.

11.1.3.2 CUT Second Kind HyperCosmos Introduction of Internal Symmetries

In the case of Second Kind HyperCosmos spaces and universes where we assume the parent and child spaces (for consistency) are both 2nd Kind HyperCosmos spaces, we find the relevant CUT equations are

$$\begin{aligned} N_s &= 2^{r/2 + r'/2} \\ N_{s2tot} &= 2^{r/2 + r'/2 + 6} N_s \\ &= 2^{r + 3 + r' + 3} = d_{d2r} d_{d2r'} \end{aligned} \qquad (11.30)$$

where d_{d2r} is the 2nd Kind HyperCosmos dimension array size for space-time dimension r. In this case we make each fermion creation/annihilation operator a Cayley number[78] operator of Cayley number $2^{r/2+ r'/2+ 6}$. This assumption is designed to obtain the 2nd Kind HyperCosmos form for the size of dimension arrays. Thus we have 2nd Kind HyperCosmos IPF's as well. See Fig. 11.4 for the 2nd Kind IPF dimension array sizes for parent and child spaces.

11.1.3.3 IPF's with Multiple Child Spaces

Defining an IPF with multiple child spaces/universes is possible. For example for HyperCosmos spaces a free IPF with two sibling child spaces/universes is.

$$\begin{aligned} \psi_{i\alpha\beta\gamma}(y, z, x) = \sum_{s_1} \sum_{s_2} \sum_{s_3} \int dp^{r-1} \int dq^{r'-1} \int dk^{r''-1} N(p, q, k) \, [b_i(p, q, k, s_1, s_2, s_3) u_\alpha(p,s_1) u_\beta(q, s_2) \cdot \\ \cdot u_\gamma(k, s_3) \exp(-ip \cdot y - iq \cdot z - ik \cdot x) + \\ + d_i^\dagger(p, q, k, s_1, s_2, s_3) \, v_\alpha(p,s_1) v_\beta(q, s_2) \, v_\gamma(k, s_3) \exp(ip \cdot y + iq \cdot z + ik \cdot x)] \end{aligned} \qquad (11.31)$$

plus Hermitean conjugates for i = 1, 2 where N(p, q, k) is a normalization factor.

Following a similar discussion to section 11.1.3.1 we find

$$\begin{aligned} N_{stot} &= 2^{r + 4 + r' + 4 + r'' + 4} \\ &= d_{dr} d_{dr'} d_{dr''} \end{aligned} \qquad (11.31a)$$

where d_{dr} is the dimension array size for space-time dimension r.

This dual child wave function expresses the parent-child-child relation through the form of the creation/annihilation operators in eq. 11.31. Any universe may have any number of child universes using wave function generalizations of eq. 11.31.

The above discussion applies with simple changes to sets of Second Kind HyperCosmos based universes and mixed sets of HyperCosmos and Second Kind HyperCosmos spaces. For three Second Kind HyperCosmos spaces in a parent-child-child relation:

---

[78] The Cayley numbers have the values $C_n = 2^n$.

$$N_{s2tot} = 2^{r+3+r'+3+r''+3} = d_{d2r}d_{d2r'}d_{d2r''} \tag{11.31b}$$

11.1.3.4 IPF for a Three Generation set of Universes

To define a wave function for a parent-child-grandchild set of universes we use two sets of PseudoQuantum wave functions. The rationale for this choice is the bond between a parent and child at any level of descent. A child universe only exists within a single parent universe.[79]

Parent-Child

$$\psi_{i\alpha\beta}(y, z) = \sum_{s_1} \sum_{s_2} \int dp^{r-1} \int dq^{r'-1} N(p, q) [b_i(p, q, s_1, s_2)u_\alpha(p,s_1)u_\beta(q, s_2) \exp(-ip\cdot y - iq\cdot z) + $$
$$+ d_i^\dagger(p, q, s1, s2)v_\alpha(p,s1)v_\beta(q, s2) \exp(ip\cdot y + iq\cdot z)] \tag{I}$$

plus Hermitean conjugates for i = 1, 2 where N(p, q) is a normalization factor.

Child-Grandchild

$$\psi_{i\alpha\beta}(z, w) = \sum_{s_1} \sum_{s_2} \int dp^{r'-1} \int dq^{r''-1} N(p, q) [b_i(p, q, s_1, s_2)u_\alpha(p,s_1)u_\beta(q, s_2) \exp(-ip\cdot z - iq\cdot w) + $$
$$+ d_i^\dagger(p, q, s1, s2)v_\alpha(p,s1)v_\beta(q, s2) \exp(ip\cdot z + iq\cdot w)] \tag{II}$$

plus Hermitean conjugates for i = 1, 2 where N(p, q) is a normalization factor.[80]

The separation of parent-child and child-grandchild creation/annihilation operators between I and II is required to separate the generations.

The dimension array discussion for three generations of universes starts with the spin sums:

$$N_{SpinsParent\text{-}Child} = 2^{r/2 + r'/2}$$
$$N_{SpinsChild\text{-}Grandchild} = 2^{r'/2 + r''/2}$$

Each of the sets of parent-child and child-grandchild operators has its operators acquire a Cayley number index generating the factors $2^{r/2 + r'/2 + 4}$ and $2^{r'/2 + r''/2 + 4}$ respectively[81] with the results:

$$N_{stotParent\text{-}Child} = 2^{r+4+r'+4} = d_{dr}d_{dr'}$$
$$N_{stotChild\text{-}Grandchild} = 2^{r'+4+r''+4} = d_{dr'}d_{dr''}$$

The spaces of all three universes conform to the HyperCosmos spaces dimension arrays sizes. This approach may be used to generate multigeneration sequences of universes.

---

[79] The possibility of a child universe with two or more parents is considered in section 11.1.3.5.

[80] The repetition of the summation variables in I and II is acceptable since they are summed over.

[8181] Note the child multiplicative factor is the same for parent-child and child-grandchild cases.

The above discussion applies with simple changes to sets of Second Kind HyperCosmos based universes and mixed sets of HyperCosmos and Second Kind HyperCosmos spaces.

11.1.3.5 Creation of Hierarchies and Networks of Universes

The preceding subsections show how to create a hierarchy[82] of universes starting from one universe. (We chose the r = 18 space and universe in earlier discussions.) Each universe has a single parent universe, any number of sibling universes, and any number of child universes.[83] We believe the Physical Cosmos is hierarchical. See Fig. 11.5 for a universe hierarchy example.

It is possible but unlikely in the author's opinion, that our Physical set of universes may form a network. In a network scenario if one selects a universe, then following a chain of universes one may come back to the starting place universe. For example a network universe may have the sequence of universes of space-times 12 10 8 6 8 12. One starts from 12 and returns to it. The 8 and 6 space-time dimension universes may be different going down the path and going up the path. For a network of universes one must have cases where a lower dimension universe is parent to a larger dimension universe.

[82] In all these discussions we treat universes as initially flat and we treat the metric as classical – c-numbers.
[83] Universes occur in hierarchies (or networks). Spaces occur within spectrums.

THE HYPERCOSMOS SPACE SPECTRUM

Blaha Space Number $N = o_s$	Cayley-Dickson Number n	Cayley Number Cn d_c	Dimension Array column length d_{cd}	Dimension Array Size d_{dN}	Space-time-Dimension r	CASe Group $su(2^{r/2},2^{r/2})$ CASe
0	10	1024	2048	2048^2	18	su(512,512)
1	9	512	1024	1024^2	16	su(256,256)
2	8	256	512	512^2	14	su(128,128)
3	7	128	256	256^2	12	su(64,64)
4	6	64	128	128^2	10	su(32,32)
5	5	32	64	64^2	8	su(16,16)
6	4	16	32	32^2	6	su(8,8)
7	**3**	**8**	**16**	$\mathbf{16^2}$	**4**	**su(4,4)**
8	2	4	8	8^2	2	su(2,2)
9	1	2	4	4^2	0	su(1,1)
10	0	1	2	2^2	-2	
11	-1	½	1	1	-4	

Figure 11.1. The CUT HyperCosmos space spectrum augmented with N = 10 and N = 11 lines for use later. (Spaces with negative space-times may have universes.)

HYPERCOSMOS OF THE SECOND KIND SPACE SPECTRUM

Blaha Space Number $N = O_s$	Cayley-Dickson Number n	Cayley Number d_c	Dimension Array size d_{dN2}	Space-time-Dimension r	CASe Group $su(2^{r/2},2^{r/2})$ CASe
0	10	1024	1024 × 2048	18	su(512,512)
1	9	512	512 × 1024	16	su(256,256)
2	8	256	256 × 512	14	su(128,128)
3	7	128	128 × 256	12	su(64,64)
4	6	64	64 × 128	10	su(32,32)
5	5	32	32 × 64	8	su(16,16)
6	4	16	16 × 32	6	su(8,8)
7	**3**	**8**	**8 ×16**	**4**	**su(4,4)**
8	**2**	4	4 × 8	2	su(2,2)
9	**1**	2	2 × 4	0	su(1,1)
10	0	1	1 × 2	-2	
11	-2	½	½	-4	

Figure 11.2. The HyperCosmos of the Second Kind space spectrum augmented with N = 10 and N = 11 lines for use later.

N	r	Physically allowed r' values	d_{dN}	$d_{dN'}$' of Universes
0	18	4, 6, 8, 10, 12, 14, 16	2^{22}	2^8 2^{10} 2^{12} 2^{14} 2^{16} 2^{18} 2^{20}
1	16	4, 6, 8, 10, 12, 14	2^{20}	2^8 2^{10} 2^{12} 2^{14} 2^{16} 2^{18}
2	14	4, 6, 8, 10, 12	2^{18}	2^8 2^{10} 2^{12} 2^{14} 2^{16}
3	12	4, 6, 8, 10	2^{16}	2^8 2^{10} 2^{12} 2^{14}
4	10	4, 6, 8	2^{14}	2^8 2^{10} 2^{12}
5	8	4, 6	2^{12}	2^8 2^{10}
6	6	4	2^{10}	2^8
7	4	–	2^8	-

Figure 11.3 The spectrum of r' values for the HyperCosmos space-times r of CUT. The dimension array sizes for the r space-time are d_{dN}. The dimension array sizes for the r space-time are $d_{dN'}$' and are lised in the 5th column for r''s in the 3rd column.

N	r	Physically allowed r' values	d_{dN}	$d_{dN'}$' of Universes
0	18	4, 6, 8, 10, 12, 14, 16	2^{21}	2^7 2^9 2^{11} 2^{13} 2^{15} 2^{17} 2^{19}
1	16	4, 6, 8, 10, 12, 14	2^{19}	2^7 2^9 2^{11} 2^{13} 2^{15} 2^{17}
2	14	4, 6, 8, 10, 12	2^{17}	2^7 2^9 2^{11} 2^{13} 2^{15}
3	12	4, 6, 8, 10	2^{15}	2^7 2^9 2^{11} 2^{13}
4	10	4, 6, 8	2^{13}	2^7 2^9 2^{11}
5	8	4, 6	2^{11}	2^7 2^9
6	6	4	2^9	2^7
7	4	–	2^7	-

Figure 11.4 The spectrum of r' values for the 2nd Kind HyperCosmos space-times r of CUT. The dimension array sizes for the r space-time are d_{dN}. The dimension array sizes for the r space-time are $d_{dN'}$' and are lised in the 5th column for r' values in the 3rd column.

A Hierarchical Cosmos

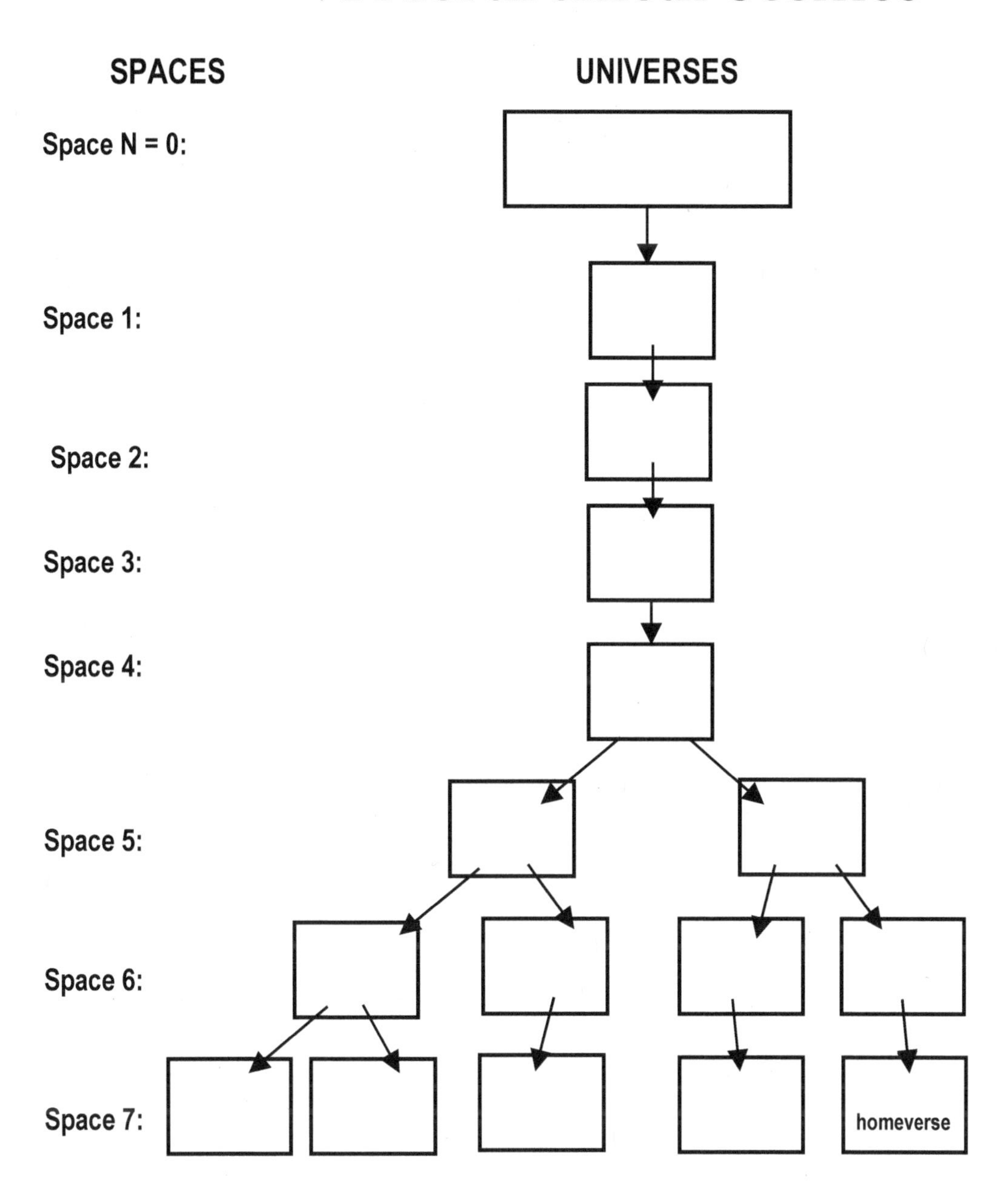

Figure 11.5. A hierarchy of universes (and possibly anti-universes) leading from the N = 0 space to the "homeverse" – our designation for our N = 7 universe. The homeverse has one "sibling" and three "cousin" universes. *The entire hierarchy resides in the N = 0 universe since the inheritance stems from the N = 0 universe. Other universes could be "reached" from the N = 0 universe if a mode of transportation exists.*

11.2 Lowest "Energy" CUT Universe States

The CUT ProtoCosmos "atom" with the "lowest" energy" and thus the highest space-time dimension has r = 18. The dimension array d_{d18} contains $2^{18+4} = 2^{22}$ elements. It has two PseudoQuantum wave functions:

$$\psi_{i\alpha\beta}(y, z) = \sum_{s_1} \sum_{s_2} \int dp^{17} \int dq^{r'-1}\, N(p, q)\, [b_i(p, q, s_1, s_2)u_\alpha(p,s_1)u_\beta(q, s_2) \exp(-ip\cdot y - iq\cdot z) +$$

$$+ d_i^\dagger(p, q, s_1, s_2)v_\alpha(p,s_1)v_\beta(q, s_2) \exp(ip\cdot y + iq\cdot z)] \qquad (11.32)$$

for i = 1, 2.

As we have done previously in models reaching back to the 1970s we choose the i = 2 creation/annihilation operators to define universe states. Thus

$$b_2^\dagger(p, q, s_1, s_2)|0> \qquad (11.33)$$

defines parent and child universes of mass M in r space-time dimensions and r' space-time dimensions respectively. The state specifies the parent universe with momentum p in the r = 18 dimension parent universe. The state specifies the child universe with momentu q in the r' dimension universe. The spins are s_1 and s_2. We may consider the special case of a universe in its rest frame where $q^0 = M$ and $\mathbf{q} = \mathbf{0}$. We take the initial universe to be concentrated in a sphere of radius 1/M. Then, within this infinitesimal universe, we assume there is a massive amount of mass-energy that causes an expansion from an initial Big Bang state.[84]

We assume both parent and child universes are flat and infinite initially. Dynamical evolution may make either universe curved.

The child universe viewed as a particle[85] has a size of the order of 1/M. We assume M is very large and constitutes the mass-energy of the child universe at the point of creation.

Superpositions in the parent space can be used to localize the universe particle in the parent space. For example,

$$\int dp^{r-1}\, f(p)\, b_2^\dagger(p, q = (M, \mathbf{0}), s_1, s_2)|0> \qquad (11.34)$$

generates a superposition in the parent space specified by f(p) with one child universe at rest.

11.2.1 Gestation of a Universe

A universe may be created directly from the CUT eigensolutions. A universe may be created through an interaction. We can consider a QED-like interaction that leads to a "photon" splitting into a universe-antiuniverse pair. In previous books we

---

[84] See Blaha (2004) and (2019e) for universe expansion from a Big Bang state. This derivation appears to be the only detailed derivation of Big Bang dynamics.
[85] See Blaha (2021d).

suggested that universes have a "charge." A universe charge may be taken to be +1; an antiuniverse charge would then be –1. Universes may also be created in universe-universe interactions. See section 11.3 below.

In an "instant" after creation a universe begins to expand dramatically. The initial state is an ultracompact particle of mass-energy equipartitioned since all interactions are "zero" and all fermions and bosons are massless – a soup of mass-energy. After the universe state is created the dense energy-matter content causes a Big Bang that leads to Hubble-like expansion of the universe.[86]

It then rapidly expands Hubble-wise with particles acquiring mass and interactions. Then one sees the growth of stars and galaxies.

11.2.2 HubbaHubble Universe Expansion

The Hubble scale factor specifies the "size" of a universe as a function of time. We have specified a fit to the Hubble expansion, H(t), of our universe in Blaha (2021d) that accounts for the known aspects of the Hubble expansion. We call it the *HubbaHubble model*:

$$a_{HH}(t) = \underbrace{[(t + t_0)/t_{now}]^{gd/(t + t_0)}}_{\textbf{I}}\underbrace{[(t + t_0)/t_{now}]^{g}}_{\textbf{II}}\ \underbrace{[(t + t_0)/t_{now}]^{h(t + t_0)}}_{\textbf{III}} \tag{11.35}$$

where

$$da_{HH}(t)/dt = H(t):$$

It has three parts that apply almost independently to stages of universe expansion.

IF an exact calculation were performed using "known" H(t) values

$$H(380{,}000\text{ yr}) = 67.8 \tag{11.36}$$
$$H(t_{now}) = 73.24 \tag{11.37}$$

then eqs. 11.36 and 11.37 with

$$h = (t_c H(t_c) - t_{now}H(t_{now}))[\, t_c - t_{now} + t_c \ln((t_c/t_{now})]^{-1} \tag{11.38}$$
$$g = (H(t_{now}) - h)\, t_{now}$$

imply[87]

$$h = 2.25983 \times 10^{-18} \tag{11.39}$$
$$d = 2.956 \times 10^{-194}\text{ s} \tag{11.40}$$
$$g = 0.000282377 = 2.82377 \times 10^{-4} \tag{11.41}$$

Note:

$$h \cong 1/t_{now} \tag{11.42}$$

[86] See Blaha (2004) and (2019e). This derivation appears to be the only detailed derivation of Big Bang dynamics.

[87] **The calculation of g, in particular, is delicate since it contains small differences between large quantities.**

11.2.2.1 Physial Understanding of a_{HH}

The parts of a_{HH} may be viewed from the perspective of particle Physics if one examines the Fourier transforms of the three parts.

Region I

The region I $a_{HH}(t)$ is approximately

$$a_{HHI}(t) \;=\; [(t + t_0)/t_{now}]^{gd/(t + t_0)} \tag{11.43}$$

This region may be viewed as a subregion where $0 \le t \le t_0$, and as another subregion where $-t_0 < t < 0$. In the first subregion, which is part of the Big Bang region, we see

$$a_{HHI}(t) \;\cong\; [t_0/t_{now}]^{gd/\,t_0} \tag{11.44}$$

is constant to good approximation. Note $a_{HH}(0)$ is a finite, non-zero number.

In the second subregion, which is also part of the Big Bang region, $a_{HH}(t)$ varies up to an infinite value at $t = -t_0$. We view the divergence, an essential singularity, as an indication of a transition from a fermion-antifermion pair to a universe particle in HyperCosmos Cosmology. Its analog is an essential singularity in the ultraviolet region in the vacuum polarization that was conjectured by Adler.[88] The $a_{HHI}(t)$ behavior may be seen in the extreme left of the plot.

Region II

The region II behavior is similar to vacuum polarization. See Blaha (2021d).

Region III

The region III $a_{HH}(t)$ behavior is approximately

$$\begin{aligned} a_{HHIII}(t) \;&=\; [(t + t_0)/t_{now}]^{g\,+\,h(t + t_0)} \\ &\cong\; [t/t_{now}]^{g\,+\,ht} \end{aligned} \tag{11.45}$$

It is governed primarily by the “ht” exponent. The implicit logarithms in the $t^{ht} = e^{ht\ln t}$ factor suggest that the corresponding Fourier transform has a vacuum polarization-like infrared part similar to the soft photon-like behavior seen in QED.

We conclude the behavior of $a_{HH}(t)$ is similar to a Fourier transformed particle QED-like vacuum polarization.

11.2.2.2 Consequences of the HubbaHubble $a_{HH}(t)$ Model

The $a_{HH}(t)$ model has an impressive set of features:

---

[88] S. Adler, Phys. Rev. **D5**, 3021 (1972).

1. The scale factor becomes very small, but non-zero, as $t \rightarrow 0$, the Big Bang point. The scale factor $a_{HH}(t)$ is constant in the Big Bang region from $t = 0$ to 10^{-200} s. The essential singularity part with the parameter d greatly affects a and H in the Big Bang region.

2. $H = a_{HH}(t)$ becomes very large as $t \rightarrow 0$.

3. The large time features of $a_{HH}(t)$ are consistent with known data.

4. From t =o to $t = 10^{-200}$ s $a_{HH}(t)$ is flat. This region is the Big Bang region.

5. The constants g and h are determined from H(t) in the large t region. The parameters d and t_0 are determined by $a_{HH}(t)$, H and g in the Big Bang region.

6. The essential singularity at $t = -t_0$ *may* represent the point of a fermion-antifermion annihilation, or of "photoproduction" of a universe-antiuniverse pair to create universes.

Thus we have a satisfactory set of parameters for the HubbaHubble model extending to the Big Bang period. The additional factor with an essential singularity in time yields a satisfactory model for t = 0 to the present.

11.2.2.3 A Big Dip in H(t) and $a_{HH}(t)$

Fig. 11.6 shows a Big Dip first suggested in Blaha (2021d). This feature is to be expected since H(t) declines from a large positive value at small times and rises at large times near the present. ***A minimum must be present by simple algebra.***

The Big Dip events took place at:

$$\text{Big Dip minimum } (H \cong -410) \text{ at } t = t = \; 4.1199 \times 10^{14} \text{ sec.}$$

Since the changeover from a radiation-dominated phase to a matter-dominated phase is approximately estimated to be at:

$$\text{Radiation – Matter Domination Transition: } t \cong 1.48 \times 10^{12} \text{ sec.}$$

Due to experimental uncertainties it may coincide with the Big Dip.

It seems reasonable to conclude the transition from radiation-dominated to matter-dominated causes the Big Dip to occur. The matter-dominated phase transition causes shrinkage. *The universe contracts by one-third!* [89] We attribute the time delay between the transition and the Big Dip in $a_{HH}(t)$ to the time required for the transition to occur. (The universe is large at that time after all)

---

[89] Rather like the condensation of water vapor to liquid.

11.2.2.4 Universe Contraction – Early Massive Galaxies

The contraction would appear to "squeeze" the mass-energy in the universe giving it a "belly" (a Big Belly? Of "squeezed" mass-energy).This mass-energy contraction leads to the early formation of galaxies that disperse due to gravitation in the 13.5 Gyrs that follow. The subsequent expansion would also appear to create a "wake" similar to the wake of a water wave.

Evidence[90] has been found for the existence of a huge population of very massive galaxies (39+ have been found so far) that were created within one billion years after the Big Bang. This population of early galaxies is inconsistent with the standard present-day models of galaxy formation. The Big Dip occurs at 2.76 million years – well before one billion years – consistent with the formation of early massive galaxies.

A concentration of mass-energy due to the contraction of the universe appears to present a possible solution. Universe contraction was not considered in the creation of models of galaxy formation.

Another possible source of universe concentration (and voids) of energy appears in our Quantum Big Bang Model. The cause is a large difference in expansion rates (Hubble Constant variations) at the center of the Big Bang compared to the outer edge of the Big Bang.

11.2.2.5 Overshoot in H(t)

The result of the Radiation-Matter transition seems to be a negative H(t) for the energy density Ω_T. H(t) "overshoots" and becomes negative. Crudely put, the clumping of matter in the matter-dominated phase appears to introduce a compactness that results in a decrease in universe size and concentrations of energy.

11.2.2.6 Voids and Bubbles in Space after the Big Dip

The Big Dip concentrates mass-energy at the contraction. The subsequent expansion creates a "type" of wave that generates massive galaxies (bubbles of mass-energy), and also voids – bubbles of space devoid of galaxies. In the course of the following thirteen or so billion years gravitation causes a dispersion of galaxies, voids and bubbles leading to the present day observed distribution.

11.2.2.7 Mystery of the Big Dip in H(t) - A Scenario

At the Big Dip H(t) changes from a declining to a rising trajectory. Based on this fact and the Big Bang model presented in Blaha (2019c) the following scenario seems reasonable:

1. The initial peak, and immediate decline, in H(t) is due to the Y black body radiation phase pressure that decreases rapidly after the Big Bang metastate

[90] T. Wang *et al,* Nature **572**, 211 (2019).

ends. Thus Ω_T declines rapidly with the Y pressure decline. (Note Ω_T is a sum of energy density and pressure.)

2. The peak in Ω_T reflects an influx of energy (from the Megaverse?) that causes H(t) to begin increasing. There is also a dip below zero in H(t) signifying the shrinkage of the universe as a(t).

3. Afterwards Ω_T continues to be significant and increasing as a(t) and H(t) rise to the present time.

4. In the future Ω_T should continue to rise. The energy increase that this situation implies suggests a certain reality to our HubbaHubble model.

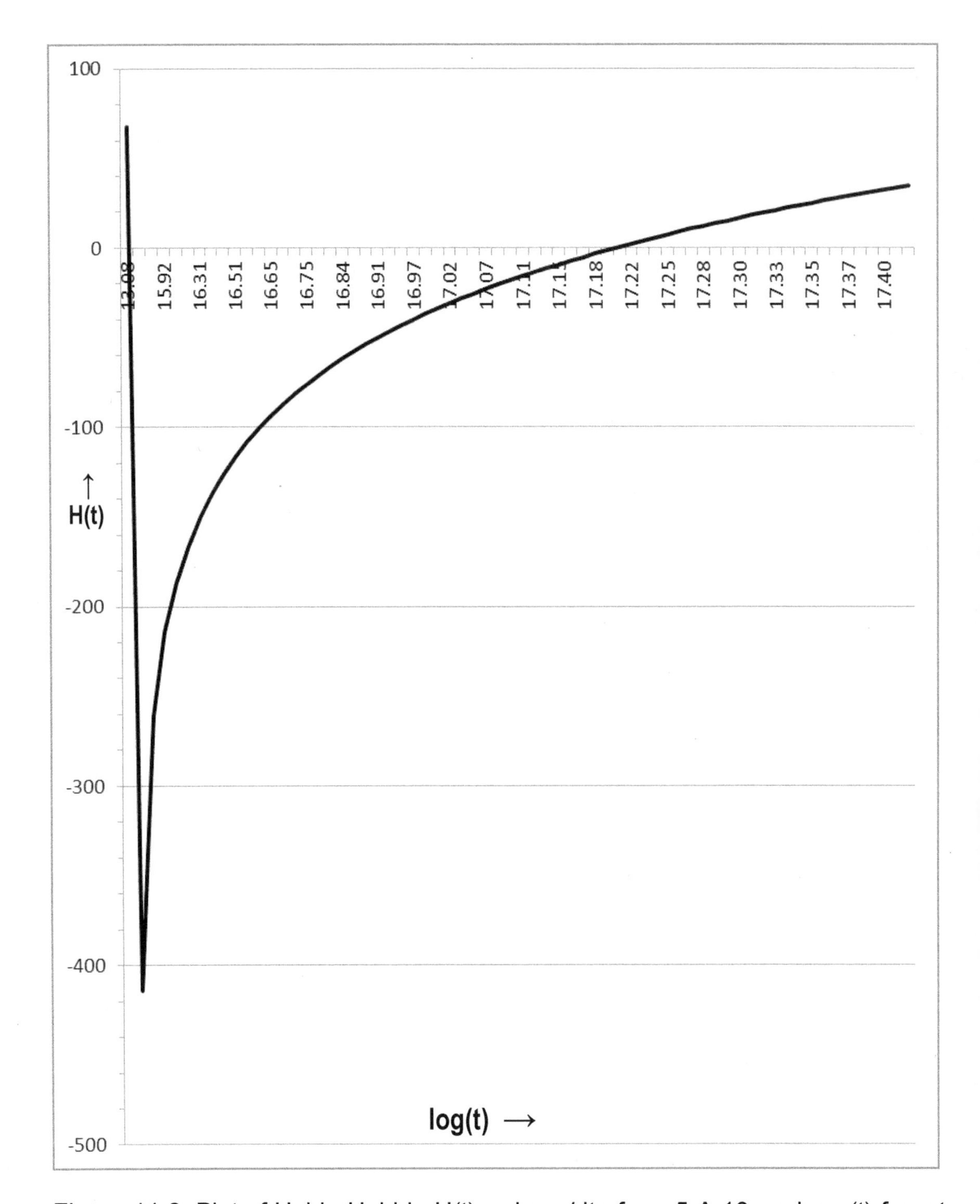

Figure 11.6. Plot of HubbaHubble $H(t) = da_{HH}/dt$ of eq. 5-A.10 vs. $\log_{10}(t)$ from t = 1.198×10^{13} to t = 5.08×10^{17} s. The Big Dip, the minimum of H(t), occurs at t = 4.1199×10^{14} s "shortly" after the radiation–matter transition in the universe

11.3 Universe Generation from Universe Interactions

A space must be distinguished from a universe. A space provides a template or blueprint, which specifies the structure and the form of the contents of a universe. A universe is a state created from a set of quantum states supported by a quantum wave function. We chose to use the IPF form of wave functions since they neatly separate a parent space universe from its child space universe(s).

We have seen how the template for a universe is responsible for the form of the Internal Symmetry groups, the space-time, and the fermion and boson particle spectrums. The created forms of universes undergo dynamical evolution according to Physical dynamic laws including symmetry breakdown, Higgs Mechanism etc.

Given a universe such as the 18 space-time dimension lowest ProtoCosmos space and universe, we now consider "sub-universe" creation.

We see three mechanisms for universe creation: spontaneous universe-antiuniverse creation from an unstable vacuum state; creation from a QED-like photon decaying into universe-antiuniverse pairs as part of a larger process; and creation of universe-antiuniverse pairs due to the interaction of a pair of universes through a universe-universe interaction that may be QED-like.

11.4 Possible Experimental Evidence for Other Universes

At first glance it would seem impossible to produce evidence for the existence of other universes. However there are subtle means by which we can 'sense' experimentally 'nearby' universes should they exist. The mechanism would appear to be gravitational effects exerted on objects within our universe by unseen objects of enormous mass. Currently there appears to be three experimental suggestions of the existence of 'nearby' universes and one theoretical argument based on an influx of mass-energy from the Megaverse that may cause the expansion of our universe.

11.4.1 Great Attractors

One potential support is the discovery of the Great Attractor (at the center of the Laniakea Galaxy Supercluster), and the more massive Shapley Attractor (centered in the Shapley Supercluster)[91]. These attractors contain massive numbers of galaxies and are drawing galaxies over a distance of millions of light years towards them.

If another universe(s) is 'near' our universe it could act as a 'gravitational magnet' and draw galaxies within our universe towards it to form one or more superclusters which could then act as attractors. Thus attractors might indirectly reveal the presence of other nearby universes—contrary to the expected large scale uniformity of the universe. The only other apparent source of superclusters is chance. Chance seems an unsatisfactory possibility in the present case.

11.4.2 Bright Bumps in Universe Suggesting Collision with Another Universe

A recent study[92] of the residual brightness of parts of the accessible universe found that bright patches appeared if a model of the CMB (Cosmic Microwave

[91] Tully, R. Brent; Courtois, Helene; Hoffman, Yehuda; Pomarède, Daniel, "The Laniakea Supercluster of galaxies". Nature (4 September 2014). 513 (7516): 71–73; arXiv:1409.0880.
[92] Ranga-Ram Chary, arXiv.org:/1510.00126 (2015).

Background) with gases, stars and dust was 'subtracted' from the PLANCK map of the entire sky. After the subtraction one would expect only noise spread throughout the sky. However, bright patches were seen in a certain range of frequencies. These anomalies are thought to be a result of our universe colliding with another object – presumably another universe in the Megaverse.

11.4.3 Cold Spot in Universe Suggesting Collision with Another Universe

Another recent study[93] of a huge cold region of the universe spanning billions of light years revealed that this region is not a relatively empty region but rather is similar to in its distribution of galaxies to the rest of the universe. Previous the Cold Spot (an area where cosmic microwave background radiation – the leftover Big Bang radiation is weak – making it significantly colder (0.00015C colder) than the average temperature of the universe.)

An analysis of 7,000 galaxy redshifts using new high-resolution data has now shown that the Cold Spot is similar to the rest of the universe. The Durham University group suggested that the Cold Spot might have been caused by a collision between our universe and another Universe. They further suggested that there is only a 1 in 50 chance that it could explain by standard cosmology.

Thus we have another important piece of circumstantial evidence in favor of other universes and thus the Megaverse.

11.4.4 Megaverse Energy-Matter Infusion into Our Universe

In chapter 14 of Blaha (2017c) we presented a model for an influx of mass-energy from the Megaverse to support the Bond-Gold-Hoyle-Narlikar Steady State Cosmology, which was originally based on the 'continuous creation of mass-energy' by Hoyle and Narlikar. This model explains why the value of Ω makes the universe close to flat. If this model is correct then we would have concrete support for a Megaverse with a low mass-energy density leaking mass-energy into our universe. *More generally, it suggests that universes are surfaces of high mass-energy density in a Megaverse of low mass-energy density – with a ratio of mass-energy densities of the other of 10^{30}.*

11.4.5 Conclusion

We conclude that data is beginning to emerge favoring multiple universes and a physical Megaverse in support of the theoretical justifications presented earlier.

---

93 T. Shanks et al, Durham University (Australia), Monthly Notices of the Royal Astronomical Society, 2016 .

12. Dimension Array Algebra and Generation

12.1 Role of HyperUnification Spaces

The general role of HyperUnification is to take the dimension array of a space and to define a "larger" space whose General Relativistic (GR) transformations mix the elements of the space's dimension array treated as a vector. In doing this we note that the dimension array of a space of space-time dimension r contains a space-time part,[94] which is also of dimension r, and an Internal Symmetries part with the dimensions of the irreducible representations of its symmetries.

The elements of the dimension array of a space are mapped to the elements of the column vector of the dimension array of the corresponding HyperUnification space. In general a GR transformation of a HyperUnification space mixes all HyperUnification space dimension array elements and, thereby, mixes all elements of the HyperCosmos space's entire dimension array. The complete mixing implements unification of space-time and Internal Symmetries.

The PseudoFemion discussions in chapter 4 of Blaha (2023c) showed that PseudoFermions of space-time dimension r in the ProtoCosmos may be viewed as existing in a subspace of a HyperUnification space of space-time dimension 2r + 4. Thus they are naturally unified under HyperUnification space GR transformations.

We now turn to consider the three forms of unification mixing in Cosmos Theory. Later we will develop dimension array algebra and describe a generation mechanism for HyperCosmos and Second Kind HyperCosmos spaces.[95]

12.1.1 HyperUnification of One HyperCosmos Space

A HyperCosmos space of space-time dimension r has a HyperUnification space with space-time dimension 2r + 4. The dimension array of the HyperCosmos space contains 2^{r+4} dimensions. Its HyperUnification space has a dimension array with column vectors of 2^{r+4} elements, one column of which may be taken to be the elements of the HyperCosmos space's dimension array. A purely GR transformation of the HyperUnification dimension array mixes the elements of a column vector, which is then a mixing of all the elements of the HyperCosmos space's dimension array—thus achieving a unification of its space-time and Internal Symmetries.

12.1.2 HyperUnification of One Second Kind HyperCosmos Space

A Second Kind HyperCosmos space of space-time dimension r has a HyperUnification space with space-time dimension 2r + 4. The dimension array of the space contains 2^{r+3} dimensions. Its HyperUnification space has a dimension array with column vectors of 2^{r+3} elements, one column of which may be taken to be the elements of the Second Kind HyperCosmos space's dimension array. A purely GR

[94] The space-time dimensions are taken to be real-valued—not hypercomplex numbers.
[9595] Abstracted in part from Blaha (20232c).

transformation of the HyperUnification space dimension array mixes the elements of a column vector, which is then a mixing of all the elements of the Second Kind HyperCosmos space dimension array—thus achieving a unification of its space-time and Internal Symmetries.

12.1.3 Full HyperUnification of the Sum of HyperCosmos and Second Kind HyperCosmos Spaces

One may form a square composite array containing the 20 HyperCosmos and Second Kind HyperCosmos spaces' dimension arrays arranged along the diagonal plus an 8 by 8 block of dimensions (as seen in chapter 11.) This square array becomes the dimension array of the Full Unification space with space-time dimension 42.[96]

This 42 space-time dimension array has a set of purely GR transformations that mix all the parts of the composite dimension arrays achieving a new form of unification of all 20 HyperCosmos and Second Kind HyperCosmos spaces. The major benefit of this formulation is the capability[97] it gives of generating all elements of all dimension arrays from one element using a general GR transformation.

12.1.4 HyperUnification of the 42 Dimension Full HyperUnification Space in the 88 Dimension UltraUnification Space

The Full HyperUnification space of 42 space-time dimensions has a dimension array containing 42 space-time elements (real-valued coordinates) plus 2^{46} minus 42 elements that furnish the irreducible representation dimensions of its Internal Symmetry groups.

Again it is possible to unify the dimension array of this space in an UltraUnification space[98] of 88 space-time dimensions (where 2*42 + 4 = 88). Purely GR transformations of the 88 space-time square dimension array column vector mix all elements of the 42 dimension space-time's dimension array providing complete unification.

Thus one may generate all elements of all Cosmos theory spaces from one UltraUnification space dimension array element thus furnishing an ultimate Beginning conceptually.

12.2 Dimension Array Algebra for HyperUnification Spaces

There are several Cayley number identities that are relevant for Cosmos Theory. One identity relates to the spaces spectrum of the HyperCosmos and of the Second Kind HyperCosmos spaces. Another identity is of direct importance for the 42 space-time dimension Full Unification Space.

12.2.1 HyperCosmos Dimension Array Columns and Sizes

The nth Cayley-Dickson number Cayley number C_n is

$$C_n = 2^n \tag{12.1}$$

---

[96] See chapter 6 of Blaha (2022a).
[97] See chapter 7 of Blaha (2022a).
[98] See chapter 9 of Blaha (2022a).

The corresponding number of elements in a dimension array column vector is d_{cdn} for a *HyperCosmos* dimension array where

$$d_{cdn} = 2C_n \tag{12.2}$$

The Nth HyperCosmos space dimension array (Fig. 6.1) size is

$$d_{dN} = (2C_n)^2 \tag{12.3}$$

where Blaha number N satisfies

$$N = 10 - n \tag{12.4}$$

We express d_{dN} with

$$d_{dn} = d_{cdn}^{\ 2} = (2C_n)^2 = 2^{2n+2} \tag{12.5}$$

and

$$d_{dn} = 2^{r+4} \tag{12.5a}$$

which implies

$$r = 2n - 2 \tag{12.5b}$$

Noting

$$S = \sum_{i=0}^{k} C_i = C_{k+1} - 1 \tag{12.5c}$$

we see the sum of d_{cdn} gives the sum of dimension array column lengths:

$$d_{cdk+1} = \sum_{i=0}^{k} d_{cdi} + 2 \tag{12.6a}$$

or

$$\sqrt{d_{dk+1}} = \sum_{i=0}^{k} \sqrt{d_{di}} + 2 \tag{12.6b}$$

Thus the HyperCosmos dimension array column lengths sum to the next large space's column length $d_{cdk+1} - 2$ by eq. 12.6a. The square roots of the dimension array sizes also satisfy a simple summation rule.

If we now form the sum

$$S_k = \sum_{i=0}^{k} d_{di} = 4\sum_{i=0}^{k} C_i^{\ 2} = 4 \sum_{i=0}^{k} 4^i$$

then we find a geometric sum, which yields

$$S_k = 4/3\ (4^{k+1} - 1) = 4/3\ (2^{2k+2} - 1) \tag{12.7}$$
$$= 4/3\ (d_{dk} - 1)$$

Eq. 12.7 has the first appearance of the integer 3 in our HyperCosmos formulation. This may be viewed as of some importance Philosophically. The Pythagoreans generated the integers from 1: namely 2, 3, ... The above sum takes us from the integer 2 of HyperCosmoses to the integer 3.

12.2.2 Second Kind HyperCosmos Dimension Array Columns and Sizes

The number of elements in a dimension array column vector[99] is d_{cdn} for a *Second Kind HyperCosmos* dimension array where

$$d_{cdn} = 2C_n \tag{12.8}$$

The Second Kind HyperCosmos space n[th] dimension array (Fig. 6.2) size d_{dN2} is

$$d_{d2N} = 2C_n^{\ 2} \tag{12.9}$$

where Blaha number N again satisfies

$$N = 10 - n \tag{12.10}$$

We rexpress d_{dN2} with

$$d_{d2n} = d_{cdn}^{\ 2}/2 = d_{dn}/2 = 2C_n^{\ 2} = 2^{2n+1} \tag{12.11}$$

Using

$$S = \sum_{i=0}^{k} C_i = C_{k+1} - 1$$

we see the sum of dimension array column lengths d_{cdn} is

$$d_{cdk+1} = \sum_{i=0}^{k} d_{cdi} + 2 \tag{12.12}$$

Thus the Second Kind HyperCosmos dimension array column lengths sum to the next larger space's column length – 2 by eq. 12.12.

If we now form the sum of dimension array sizes:

$$S_{2k} = \tfrac{1}{2} \sum_{i=0}^{k} d_{d2i} = 2\sum_{i=0}^{k} C_i^{\ 2} = 2 \sum_{i=0}^{k} 4^i$$

then we find a geometric sum, which yields

$$\begin{aligned} S_{2k} &= 2/3\ (4^{k+1} - 1) = 2/3\ (2^{2k+2} - 1) \\ &= 2/3(d_{d2k} - 1) \end{aligned} \tag{12.12a}$$

12.3 Consequences for the Ten HyperCosmos Spaces

The 42 space-time dimension Full HyperUnification space contains the ten HyperCosmos dimension arrays. (It also contains the ten Second Kind HyperCosmos dimension arrays, which we discuss later.) For convenient visualization we may view the ten dimension array blocks as each made into linear vector segments and then mapped to the dimension array vector of the dimension array of the 42 space-time dimension space. This mapping was discussed in detail in chapter 6 of Blaha (2023a):

---

[99] See Fig. 11.2.

$$v_S = \sum_{n=1}^{10} d_{dn} \tag{12.13}$$
$$= 4^2 + 8^2 + 16^2 + 32^2 + 64^2 + 128^2 + 256^2 + 512^2 + 1024^2 + 2048^2$$

From eq. 12.7 above we find the size of v_S

$$\begin{aligned} v_S &= S_{210} - 4 = 4/3\,(d_{d10} - 1) - 4 \\ &= 4/3(2^{22} - 1) - 4 = (2^{24} - 16)/3 \\ &= 5{,}592{,}400 \end{aligned} \tag{12.14}$$

taking account of the absence of the $d_{d0} = 4$ term.

In section 12.5 we combine the above with the similar results for the 10 Second Kind HyperCosmos spaces to complete the specification of the Full Unification Space.

12.4 Consequences for the Ten Second Kind HyperCosmos Spaces

The ten Second Kind dimension array blocks are also made into linear vector segments and then mapped to the dimension array vector of the dimension array of the 42 space-time dimension space.[100] This mapping was discussed in detail in chapter 6 of Blaha (2023a). The sum of segments now is

$$v_{S2} = \sum_{n=1}^{10} d_{dn}/2 \tag{12.15}$$
$$= [4^2 + 8^2 + 16^2 + 32^2 + 64^2 + 128^2 + 256^2 + 512^2 + 1024^2 + 2048^2]/2$$

From eq. 12.7 above we see the size of v_{S2} is

$$\begin{aligned} v_{S2} &= (S_{210} - 4)/2 = 2/3\,(d_{d10} - 1) - 2 \\ &= 2/3(2^{22} - 1) - 2 = (2^{23} - 8)/3 \\ &= 2{,}796{,}200 \end{aligned} \tag{12.16}$$

taking account of the absence of the $d_{d0} = 4$ term.

Note

$$v_{S2} = v_S/2 \tag{12.17}$$

12.5 Full Unification Space Dimension Array

The dimension array of the 42 space-time dimension space has a *column vector* with v_{42} elements:

$$\begin{aligned} v_{42} &= 2^{r/2+2} \\ &= 8{,}388{,}608 \end{aligned} \tag{12.18}$$

where r = 42. We find

$$v_{42} = v_S + v_{S2} + 8 \tag{12.19}$$

---

[100] See chapter 8 of Blaha (2023a).

exactly. The dimension array vector contains the dimension arrays of both the HyperCosmos and Second Kind HyperCosmos dimension arrays as column vector segments. Under a purely GR transformation in the 42 space-time dimension space the column vector and the segments within it are fully mixed giving a unification of all 20 component dimension arrays.

We note the sum of the HyperCosmos dimension arrays occupies 2/3 of the 42 space-time dimension array column vector and the sum of the Second Kind HyperCosmos dimension arrays occupies 1/3 of the 42 space-time dimension array column vector. Eight additional dimensions completes the 42 space-time dimension array column vector.

12.6 Addition of HyperCosmos Dimension Array Sizes

The addition of dimension array sizes is performed using eq. 12.7:

$$S_k = \sum_{i=0}^{k} d_{di} = 4/3\,(d_{dk} - 1) \qquad (12.20)$$

The difference gives

$$S_{k+2} - S_k = d_{dk+2} + d_{dk+1} = 4/3\,(d_{dk+2} - 1) - 4/3\,(d_{dk} - 1) \qquad (12.21)$$
$$= 4/3\, d_{dk+2} - 4/3\, d_{dk}$$

Upon rearranging terms we find the recurrence relation for dimension array sizes:

$$d_{dk+2} = 3\, d_{dk+1} + 4\, d_{dk} \qquad (12.22)$$

or using eq. 12.10

$$d_{d(12-N)} = 3\, d_{d(11-N)} + 4\, d_{d(10-N)} \qquad (12.23)$$

or

$$d_{dN} = 3\, d_{d(N+1)} + 4\, d_{d(N+2)} \qquad (12.24)$$

In eq. 12.22 we start with k = 0 by setting[101]

$$d_{d0} = 4$$
$$d_{d1} = 16$$

then

$$d_{d2} = 64$$
$$d_{d3} = 256$$
$$d_{d4} = 1024$$
$$\ldots$$

We may develop a closed form summation based on the recursion relation eq. 12.22. If we define

[101] See Fig. 11.1.

$$d_{dk} = x^{k+a}$$

for some integer value a and assume the recurrence relation for x we find

$$x^2 = 3x + 4$$

with the solution

$$x = 4$$

Then

$$d_{dk} = 4^{k+a}$$

for the integer value a where a = 1 in the example values d_{d0}, d_{d1}, d_{d2} , … above.

12.7 Addition of Second Kind HyperCosmos Dimension Array Sizes

Since the Second Kind dimension array sizes are half of HyperCosmos dimension array sizes

$$d_{d2n} = d_{dn}/2 \tag{12.25}$$

by eq. 12.11 we find the recursion relation are the same as eqs. 12.22 – 12.24:

$$d_{d2(k+2)} = 3\, d_{d2(k+1)} + 4\, d_{d2k} \tag{12.26}$$

$$d_{d2(12-N)} = 3\, d_{d2(11-N)} + 4\, d_{d2(10-N)} \tag{12.27}$$

or

$$d_{d2(N)} = 3\, d_{d2(N+1)} + 4\, d_{d2(N+2)} \tag{12.28}$$

In eq. 12.26 we start with k = 0 and k = 1 by setting[102]

$$d_{d20} = 2$$
$$d_{d21} = 8$$

then

$$d_{d22} = 32$$
$$d_{d23} = 128$$
$$d_{d24} = 512$$
$$\ldots$$

We may develop a closed form summation based on the recursion relation eq. 12.26. If we define

$$d_{d2k} = x^{k+½+a}$$

for some integer value a and assume the recurrence relation for x we find

$$x^2 = 3x + 4$$

with the solution

[102] See Fig. 11.2.

$$x = 4$$

Then

$$d_{d2k} = 4^{k + \frac{1}{2} + a}$$

for some integer value a where a = 0 in the example values d_{d20}, d_{d21}, d_{d22} , … above.

12.8 Multiplication of Dimension Arrays

The multiplication of the total number of elements in dimension arrays (not taking account of the values of the elements) satisfies the multiplication rules:

HyperCosmos:

$$d_{dn}d_{dn'} = d_{d(n+n'+1)} \tag{12.29}$$

Second Kind HyperCosmos:

$$d_{d2n}d_{d2n'} = d_{d2(n+n')} \tag{12.30}$$

by eqs. 12.5, 12.9, and 12.11. These rules can be used to generate the dimension arrays of the HyperCosmos and Second Kind HyperCosmos spectrums of spaces.

13. Fibonacci Numbers and Dimension Arrays

Fibonacci sequences have many applications in biology and physics. They often appear for plant and other forms of growth. They are based on summing consecutive terms. They satisfy the recurrence relation:

$$d_{k+2} = d_{k+1} + d_k \tag{13.1}$$

Some numbers in the Fibonacci sequence are 1, 1, 2, 3, 5, 8, 13, ...

In this chapter[103] we consider the relation between Fibonacci numbers and Cosmos Theory dimension arrays.

In sections 12.6 and 12.7 we derived a recurrence relation for the HyperCosmos dimension arrays sizes

$$d_{k+2} = 3\ d_{k+1} + 4\ d_k \tag{13.2}$$

and for the Second Kind HyperCosmos dimension arrays sizes:

$$d_{2(k+2)} = 3\ d_{2(k+1)} + 4\ d_{2(k)} \tag{13.3}$$

Both recurrence relations have a Fibonacci-like form. If we let $d_{-1} = 1$ and $d_0 = 4$ then we obtain a sequence of powers of 2,

$$d_k = 2^{2k+2} \tag{13.2a}$$

from eq. 13.2:, namely: 1, 4, 16, 64, 256, 1024, ... – powers of two, which are the HyperCosmos d_{dn} dimension array numbers.

If we let $d_{2\,-1} = ½$ and $d_{2\,0} = 2$ then we obtain a geometric sequence of powers of two.

$$d_{2k} = 2^{2k+1} \tag{13.3a}$$

from eq. 13.3: ½,, 2, 8, 32, 128, 512, ... which are the Second Kind HyperCosmos d_{dn2} dimension array numbers (also powers of two).[104]

These Fibonacci-like sequences show the HyperCosmos spectrums implement a form of growth—in spaces. Fig. 13.1 illustrates growth in the form of a Fibonacci-like spiral analogous to the Fibonacci spiral.

---

[103] Much of this chapter appears in Blaha (2023c).

[104] The choice of the first two numbers of a sequence is important—as the example of Lucas numbers with different starting numbers demonstrates.

13.1 Closed Power Series Summation of Fibonacci Numbers

In section 7.7 we derived the closed form summation (eq. 7.20) and the recursion relation for dimension array sizes (eq. 7.26). We may develop a similar closed form summation based on the Fibonacci recursion relation eq. 13.1. If we define

$$f_k = x^k \tag{13.4}$$

for some value x and assume the Fibonacci recurrence relation for x we find

$$x^2 = x + 1 \tag{13.5}$$

then we find the solutions

$$x = ½ \pm ½\sqrt{5} = 1.618 \tag{13.6}$$

Defining

$$\varphi = ½ + ½ \sqrt{5} \tag{13.7}$$

gives the Golden Ratio. Defining

$$\psi = ½ - ½ \sqrt{5} \tag{13.8}$$

we obtain the Fibonacci number closed form since both φ and ψ satisfy the Fibonacci recurrence relation:

$$F_k = (\varphi^k - \psi^k) /(\varphi - \psi) \tag{13.9}$$

One sees that the dimension array form of powers of 2 has a somewhat more complex analog expression for Fibonacci numbers eq. 13.9:

$$F'_k = \varphi^k/(\varphi - \psi) \tag{13.10}$$

The sum of Fibonacci numbers

$$F_{k+2} = \sum_{i=1}^{k} F_i + 1 \tag{13.11}$$

is similar to the sum of dimension array sizes:

$$d_{cdk+1} = \sum_{i=0}^{k} d_{cdi} + 2$$

Thus we see a close resemblance of Fibonacci numbers and dimension array sizes.

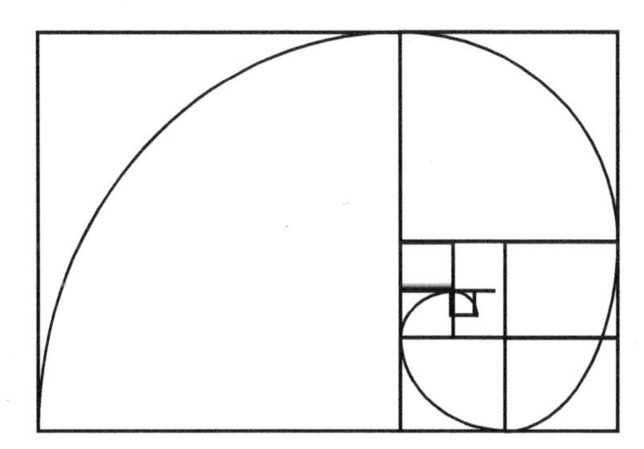

Figure 13.1. Depiction of a Fibonacci-like spiral of HyperCosmos dimension arrays using square dimension arrays. Circular arcs connecting opposing corners within the tiling.

14. Ramsey Numbers and Dimension Arrays

Ramsey Theory, and its Ramsey numbers, has wide applications in Physics and Mathematics. Ramsey Theory[105] looks for general conditions for a structure that guarantees the existence of substructures with certain regular properties. In particular, a Ramsey number is the minimum size of a group that is required to guarantee that a specific set of nodes in the group are connected to each other.

This author has defined an analogous specification within the framework of the Internal Symmetries of each Cosmos Theory space:[106]

A fermion in any block has interactions either directly, or indirectly, with every other fermion in every other block.

This criterion specifies that any fermion particle may, through a series of interactions, interact with any other fermion in the set of fermions defined for a space. Interactions reflect the Internal Symmetry group representation subsets of each space. The size of a space's dimension array, from which Internal Symmetries are defined, is a measure of the connectivity of the interactions.

In the case of our universe there are 256 fermions that have an interaction with each other directly, or through a series of interactions. This principle is part of the basis of the author's Connection Groups specification. It is needed to complete the dimension array relation to the Unified SuperStandard Theory in our universe.

Based on this similarity we use the recurrence relations of HyperCosmos dimension arrays (which are measures of the number of interactions):

$$d_{d2(k+2)} = 3\, d_{d2(k+1)} + 4\, d_{d2k} \qquad (12.26)$$

to initially suggest an analogous relation holds for Ramsey numbers:

$$R_n = R_{n-1} + 4R_{n-2} + 2[1 + (-1)^n] \qquad (14.1)$$

with the integer coefficients changed (3 replaced by 1) from eq. 7.26 above and with a term added to give the n = 4 Ramsey number the value 18. The Ramsey number recurrence relation enumeration begins with $R_1 = 1$ and $R_2 = 2$.

Eq. 14.1 generates the known four Ramsey numbers 1, 2, 6 and 18. Fig. 14.1 predicts other higher Ramsey numbers that await computation. The predicted values listed in Fig. 14.1 comfortably agree with the predicted ranges.

[105] This chapter is abstracted from Blaha (2023c).
[106] Chapter 7 of Blaha (2022f) as well as in earlier books.

14.1 Further Support for a Relation of Ramsey Numbers to Dimension Arrays

Upper and lower bounds on Ramsey numbers have been determined by Erdős and Szekeres, and Spencer and Conlon:

$$R(n,n) <= 4^{n-1}/(\pi n)^{1/2} = 2^{2n-2}/(\pi n)^{1/2} \quad (14.2)$$
$$R(n,n) >= n2^{n/2}/(2^{1/2}e) \quad (14.3)$$

The upper bound may be expressed in terms of a dimension array size d_{dn} as

$$R(n,n) <= 2^{2n-2}/(\pi n)^{1/2} = d_{dn}\,[2^{-4}/(\pi n)^{1/2}] \quad (14.4)$$

using

$$d_{dn} = d_{cdn}^{\,2} = (2C_n)^2 = 2^{2n+2} \quad (12.3)$$

The lower bound may be expressed in terms of the square root of the dimension array column size $d_{cdn}^{\,1/2}$ as

$$R(n,n) >= d_{cdn}^{\,1/2}\,[n/(2e)] \quad (14.6)$$

Thus there appears to be a connection between Ramsey numbers and Cosmos Theory dimension array sizes.

One might conjecture that the value of R(n, n) is

$$R(n, n) \sim e_n\, d_{dn}^{\,1/2} \sim e_n\, d_{cdn} \quad (14.7)$$

where e_n is a constant dependent on n and $d_{cdn} = 2^{n+1}$ is the length of the dimension array column vector. However this approximation proves to be *false* as we show in section 14.2 below.

n	Ramsey Number	Ramsey Recurrence Value using Eq. 14.1
1	1	1
2	2	2
3	6 = R(3,3)	6
4	18 = R(4,4)	18
5	43 - 48 = R(5,5)	42
6	102-161 = R(6,6)	118
7	205-497 = R(7,7)	286
8	282-1532 = R(8,8)	762
9	565-6588 = R(9,9)	1906
10	798-23556 = R(10,10)	4958

Figure 14.1. Ramsey numbers and the predicted value of Ramsey numbers from eq. 14.1. Projected ranges of Ramsey numbers are listed for n = 5 – 10.

14.2 Approximate Ramsey Numbers Related to Dimension Arrays

The recurrence relation eq. 14.1 gives the first few Ramsey number values correctly and a reasonable extrapolation of higher number values.

The recurrence relation (slightly modified) has an approximate geometric sequence form similar to that of dimension arrays. The approximate geometric sequence is

$$R_{approx}(n, n) = [2.5616^{n-1}] \tag{14.8}$$

where [] indicates greatest integer less than or equal to the quantity in brackets. Note the common ratio almost exactly (irrational numbers!) satisfies

$$2.5616 = (1 + 17^{½})/2 \cong \pi(2/3)^{½} \tag{14.8a}$$

at the 0.1% level.

Note the difference from the powers of 2 form of dimension arrays. Fig. 14.2 shows the values of this Ramsey number approximation. The approximation to R(n, n) of eq. 14.8 satisfies the following recurrence relation almost exactly:[107]

$$R_n = R_{n-1} + 4R_{n-2} \tag{14.9}$$

The algebraic equation implied by eq. 14.9 assuming $R_n = x^n$ is

$$x^2 - x - 4 = 0 \tag{14.10}$$

The solution is the common ratio x = = $(1 + 17^{½})/2 = 2.5616$.

14.2.1 Conjecture on π

Eq. 14.8a raises the possibility that the asymptotic form of the geometric sequence of Ramsey numbers may have a constant common ratio that would lead to the exact value of π.

To that end we let

$$x = (2/3)^{½}\ y$$

where y satisfies

$$2/3\ y^2 - \sqrt{2/3}\ y - 4 = 0 \tag{14.11}$$

by 14.10. Solving for y we find

$$y = \sqrt{3}\ [\sqrt{2} + \sqrt{34}]/4 = 3.137249 \cong \pi \tag{14.12}$$

a reasonable approximation to π. If there is an asymptotic common ratio x with value x = cπ for some constant c, then a new method for computing π in principle appears that is based on asymptotic Ramsey graphs.

$$\pi = x/c \doteq (3/2)^{½} x \tag{14.13}$$

---

[107] Note the similarity to eq. 7.26 above.

14.2.2 Comparison to Dimension Array Sequence

Eq. 14.7 does not successfully give approximate Ramsey numbers. The approximate Ramsey numbers are related to Cosmos Theory dimension array column sizes by

$$R_{approx}(n, n)/d_{cdn} = 2^{-2}(2.5616/2)^n = 1.2808^n/4 \qquad (14.14)$$

For[108] n = 10 we find the ratio to be 2.97. Thus $R_{approx}(10, 10)$ is $2.97d_{cd10}$ – approximately a factor of 3 – a not excessive multiple.

n	Ramsey Number	Ramsey Approximation using eq. 14.8	Ratio to Ramsey Number
1	1	[1.00] = 1	1.
2	2	[2.56] = 2	1.
3	6 = R(3,3)	[6.56] = 6	1.
4	18 = R(4,4)	[16.81] = 16	0.9
5	43 - 48 = R(5,5)	[43.01] = 43	
6	102-161 = R(6,6)	[110.3] = 110	
7	205-497 = R(7,7)	[282.5] = 282	
8	282-1532 = R(8,8)	[723.7] = 723	
9	565-6588 = R(9,9)	[1853.9] = 1853	
10	798-23556 = R(10,10)	4749	

Figure 14.2. Ramsey numbers vs. the geometric sequence predicted value of eq. 14.8.

The closeness of the approximate form to the known Ramsey numbers suggests it might be a useful guide to the calculation of the almost intractable higher n Ramsey numbers. The Ramsey number recurrence relation eq. 14.9 and approximate Ramsey number solution eq. 14.8 require further study.

The geometric sequence form of eq. 14.8 is consistent with the analogy between Ramsey numbers and dimension array numbers made earlier. The analogy may point to a deeper basis for Cosmos Theory.

It suggests a set of approximations to the large n Ramsey numbers that have proved difficult to calculate exactly.

---

[108] There are 10 HyperCosmos spaces.

Appendix A. Superluminal Quantum Field Theory

This Appendix contains two chapters on Superluminal Quantum Field Theory that was published in the early 2000's in Blaha (2007a).

2. Tachyons

A number of authors have developed field theories of tachyons within the framework of conventional Lorentz transformations. In chapter 1 of Blaha (2007a) we developed a new form of transformation, Superluminal transformations, that transform between reference frames moving with a relative speed greater than the speed of light. We then proceeded to begin the development of a theory of spin ½ tachyons and a generalized Dirac equation that is covariant under both Lorentz transformations and Superluminal transformations.

In this chapter we will discuss the canonical quantization of spin ½ tachyons on spacelike surfaces. We evade the well-known result that there are no finite dimensional representations of the Lorenz group for imaginary mass by embedding the Lorentz group within a larger group that does have finite dimensional representations for imaginary mass. This larger group, which we call the *Luminal Group*, contains the Lorentz group as an invariant subgroup. It is analyzed in chapter 6 of Blaha (2007a).

2.1 Spin ½ Tachyons

In chapter 1[109] we *provisionally* established some of the basic features of spin ½ tachyons which we list here for the reader's convenience:

$$(\gamma^\mu \partial/\partial x^\mu + m)\psi_T(x) = 0 \tag{1.89}$$

$$\mathcal{L} = \psi_T{}^S(\gamma^\mu \partial/\partial x^\mu + m)\psi_T(x) \tag{1.91}$$

$$I = \int d^4x \mathcal{L}_T \tag{1.92}$$

$$\psi_T{}^S = \psi_T{}^\dagger\, i\gamma^0\gamma^5 \tag{1.93}$$

$$\psi_T{}^S(\overleftarrow{\nabla\!\!\!/} - m) = 0 \tag{1.94}$$

$$\pi_{Ta} = \partial\mathcal{L}/\partial\dot{\psi}_{Ta} \equiv \partial\mathcal{L}/\partial(\partial\psi_{Ta}/\partial t) = -i(\psi_T{}^\dagger\gamma^5)_a \tag{1.95}$$

$$\mathcal{H} = \pi_T\dot{\psi}_T - \mathcal{L} = i\psi_T{}^\dagger\gamma^5(\boldsymbol{\alpha}\cdot\nabla + \beta m)\psi_T = -i\psi_T{}^\dagger\gamma^5\dot{\psi}_T \tag{1.96}$$

$$\{\psi_T{}^\dagger{}_a(x),\, \psi_{Tb}(x')\} = -\,[\gamma^5]_{ab}\,\delta^3(x - x') \tag{1.120}$$

[109] Equations in this Appendix that are numbered 1. … are from chapter 1 of Blaha (2007a).

$$\{\psi_{TL a}{}^{\dagger}(x), \psi_{TLb}(x')\} = ½(1 - \gamma^5)_{ab}\, \delta^3(x - x') \quad (1.121)$$

$$\{\psi_{TR a}{}^{\dagger}(x), \psi_{TRb}(x')\} = -½(1 + \gamma^5)_{ab}\, \delta^3(x - x') \quad (1.122)$$

$$\{\psi_{TL a}{}^{\dagger}(x), \psi_{TRb}(x')\} = \{\psi_{TR a}{}^{\dagger}(x), \psi_{TLb}(x')\} = 0 \quad (1.123)$$

Probability Conservation Law

The tachyon Dirac equation (eq. 1.89) implies a probability conservation law:

$$\partial\rho_5/\partial t = \nabla\cdot\mathbf{j}_5 \quad (2.1)$$

where

$$\rho_5 = \psi_T{}^{\dagger}\gamma^5\psi_T \qquad \mathbf{j}_5 = \psi_T{}^{\dagger}\gamma^5\boldsymbol{\alpha}\psi_T \quad (2.2)$$

We are thus led to define the conserved axial charge Q_5

$$Q_5 = \int d^3x\, \psi_T{}^{\dagger}\gamma^5\psi_T \quad (2.3)$$

Energy-Momentum Tensor

The tachyon energy-momentum tensor is

$$\mathfrak{T}_{T\mu\nu} = -\, g_{\mu\nu}\, \mathfrak{L}_T + \partial\mathfrak{L}_T/\partial(\partial\psi_T/\partial x_\mu)\, \partial\psi_T/\partial x^\nu \quad (2.4)$$

$$= i\psi_T{}^{\dagger}\gamma^0\gamma^5\gamma_\mu\partial\psi_T/\partial x^\nu \quad (2.5)$$

and thus the conserved energy and momentum are

$$P^0 = H = \int d^3x\, \mathfrak{T}_T{}^{00} = i\int d^3x\psi_T{}^{\dagger}\gamma^5(\boldsymbol{\alpha}\cdot\nabla + \beta m)\psi_T \quad (2.6)$$

using eq. 1.96, and

$$P^i = \int d^3x\, T_T{}^{0i} = -i \int d^3x\, \psi_T{}^{\dagger}\gamma^5\partial\psi_T/\partial\psi_i \quad (2.7)$$

Both the energy and momentum differ significantly from the corresponding quantities for conventional Dirac fields. We will look into these differences in detail later. We leave the calculation of the angular momentum expressions as an exercise for the reader.

Spin ½ Tachyon Spinors

The general form of the solutions of the free tachyon Dirac equation can be written

$$\psi_T{}^r(x) = e^{-i\chi_r p\cdot x}w^r(p) \quad (2.8)$$

where $\chi_r = +1$ for r = 1, 2 and $\chi_r = -1$ for r = 3, 4. Denoting the spinors $w^r(p) = w^r(0)$ for a particle is at rest in a frame (E = m) we see they can take the form

$$w^r(0) = \begin{bmatrix} \delta_{3r} \\ \delta_{4r} \\ \delta_{1r} \\ \delta_{2r} \end{bmatrix} \tag{2.9}$$

where Kronecker deltas appear in the brackets. The reason for the unconventional indexing is due to the transformation of positive energy Dirac spinors into negative energy tachyon spinors and vice versa (eq. 1.87). From eq. 1.86 we find[110]

$$S_S(\Lambda_S(v))w^r(0) = w_S{}^r(p) \tag{2.10}$$

Using eq. 1.77 for $S_S(\Lambda_S(v))$:

$$S_S(\Lambda_S(v)) = 2^{-\frac{1}{2}}[(\cosh(\omega/2) - i\sinh(\omega/2))I + \\ + (\sinh(\omega/2) - i\cosh(\omega/2))\gamma^0\boldsymbol{\gamma}\cdot\mathbf{p}/|\mathbf{p}|] \tag{1.77}$$

and

$$\mathbf{p} = m\mathbf{v}\gamma_s \qquad\qquad E = m\gamma_s \tag{1.81}$$

$$\sigma_- \equiv (2m)^{\frac{1}{2}}\sinh(\omega/2) = (p - m)^{\frac{1}{2}} \tag{1.82}$$

$$\sigma_+ \equiv (2m)^{\frac{1}{2}}\cosh(\omega/2) = (p + m)^{\frac{1}{2}} \tag{1.83}$$

we see that eq. 2.9 implies the columns of the resulting $S_S(\Lambda_S(v))$ matrix are

$$S_S(\Lambda_S(v)) = (2m^{\frac{1}{2}})^{-1} \begin{bmatrix} \underline{w_S{}^3(p)} & \underline{w_S{}^4(p)} & \underline{w_S{}^1(p)} & \underline{w_S{}^2(p)} \\ \sigma_+ - i\,\sigma_- & 0 & (\sigma_- - i\sigma_+)p_z/p & (\sigma_- - i\sigma_+)p_-/p \\ 0 & \sigma_+ - i\sigma_- & (\sigma_- - i\sigma_+)p_+/p & -(\sigma_- - i\sigma_+)p_z/p \\ (\sigma_- - i\sigma_+)p_z/p & (\sigma_- - i\sigma_+)p_-/p & \sigma_+ - i\sigma_- & 0 \\ (\sigma_- - i\sigma_+)p_+/p & -(\sigma_- - i\sigma_+)p_z/p & 0 & \sigma_+ - i\sigma_- \end{bmatrix} \tag{2.11}$$

with $p_\pm = p_x \pm ip_y$. It is easy to verify

---

[110] These equations are from chapter 1 of Blaha (2007a).

$$(-i\not{p} - \chi_r m)w_S{}^r(p) = 0 \tag{2.12}$$

where $\chi_r = +1$ for r = 1, 2 and $\chi_r = -1$ for r = 3, 4.

2.2 Superluminal Spinors

The spinors that we defined in eq. 2.10 can be generalized in a manner similar to Dirac spinors. We will use a similar notation to the Dirac spinor notation:

$$\begin{aligned} u_T(p, s) &= S_S(\Lambda_S(\mathbf{p}/p))w^1(0) \\ u_T(p, -s) &= S_S(\Lambda_S(\mathbf{p}/p))w^2(0) \\ v_T(p, s) &= S_S(\Lambda_S(\mathbf{p}/p))w^3(0) \\ v_T(p, -s) &= S_S(\Lambda_S(\mathbf{p}/p))w^4(0) \end{aligned} \tag{2.13}$$

where $p = |\mathbf{p}|$.

The hermitean conjugates of these spinors are

$$\begin{aligned} u_T{}^\dagger(p, s) &= w^{1T}(0)S_S{}^\dagger(\Lambda_S(\mathbf{p}/p)) \\ u_T{}^\dagger(p, -s) &= w^{2T}(0)S_S{}^\dagger(\Lambda_S(\mathbf{p}/p)) \\ v_T{}^\dagger(p, s) &= w^{3T}(0)S_S{}^\dagger(\Lambda_S(\mathbf{p}/p)) \\ v_T{}^\dagger(p, -s) &= w^{4T}(0)S_S{}^\dagger(\Lambda_S(\mathbf{p}/p)) \end{aligned} \tag{2.14}$$

where the superscript "T" indicates the transpose. We define "double dagger" spinors:

$$\begin{aligned} u_T{}^\ddagger(p, s) &= w^{1T}(0)S_S{}^\dagger(\Lambda_S(\mathbf{p}/p))i\boldsymbol{\gamma}\cdot\mathbf{p}/|\mathbf{p}| \\ u_T{}^\ddagger(p, -s) &= w^{2T}(0)S_S{}^\dagger(\Lambda_S(\mathbf{p}/p))i\boldsymbol{\gamma}\cdot\mathbf{p}/|\mathbf{p}| \\ v_T{}^\ddagger(p, s) &= w^{3T}(0)S_S{}^\dagger(\Lambda_S(\mathbf{p}/p))i\boldsymbol{\gamma}\cdot\mathbf{p}/|\mathbf{p}| \\ v_T{}^\ddagger(p, -s) &= w^{4T}(0)S_S{}^\dagger(\Lambda_S(\mathbf{p}/p))i\boldsymbol{\gamma}\cdot\mathbf{p}/|\mathbf{p}| \end{aligned} \tag{2.15}$$

.

which appear in important spinor "completeness" sums:

$$\sum_{\pm s} u_{T\alpha}(p, s)u_T{}^\ddagger{}_\beta(p, s) = (2m)^{-1}(-i\not{p} + m)_{\alpha\beta} \tag{2.16}$$

$$\sum_{\pm s} v_{T\alpha}(p, s)v_T{}^\ddagger{}_\beta(p, s) = (2m)^{-1}(-i\not{p} - m)_{\alpha\beta}$$

2.3 Second Quantization of Spin ½ Tachyon Field

In this section we will define and second quantize a tachyon field. In the case of tachyons there is, in a certain sense, a 3-dimensional time subspace since the effective metric is $-g_{\mu\nu}$. For example, the mass shell condition is $-g_{\mu\nu}p^\mu p^\nu = m^2$. Thus we must specify a "direction of time" in this 3-dimensional subspace. If we examine the form of the tachyon spinors (eq. 2.13 and 2.14) we see that we have implicitly chosen a

Superluminal "boost" for each Fourier component of the tachyon field in the direction of its 3-momentum. This is similar to the use of Lorentz boosts to define Dirac spinors as shown earlier.

Conventional Dirac Equation Solution

For the sake of comparison with tachyon field quantization we list the parts of the Fourier expansion of a Dirac field in our notation:

$$\psi(x) = \sum_{\pm s} \int d^3p\, N(p)[b(p, s)u(p, s)e^{-ip\cdot x} + d^\dagger(p, s)v(p, s)e^{ip\cdot x}] \tag{2.17}$$

$$\psi^\dagger(x) = \sum_{\pm s} \int d^3p\, N(p)[b^\dagger(p, s)u^\dagger(p,s)e^{+ip\cdot x} + d(p, s)v(p, s)e^{-ip\cdot x}] \tag{2.18}$$

where

$$N(p) = [m/((2\pi)^3 E_p)]^{1/2} \tag{2.19}$$

$$p^0 = E_p = (\mathbf{p}^2 + m^2)^{1/2} \tag{2.20}$$

where the anticommutation relations of the Fourier coefficient operators are

$$\begin{gathered}\{b(p,s), b^\dagger(p',s')\} = \delta_{ss'}\delta^3(\mathbf{p} - \mathbf{p'}) \\ \{d(p,s), d^\dagger(p',s')\} = \delta_{ss'}\delta^3(\mathbf{p} - \mathbf{p'}) \\ \{b(p,s), b(p',s')\} = \{d(p,s), d(p',s')\} = 0 \\ \{b^\dagger(p,s), b^\dagger(p',s')\} = \{d^\dagger(p,s), d^\dagger(p',s')\} = 0 \\ \{b(p,s), d^\dagger(p',s')\} = \{d(p,s), b^\dagger(p',s')\} = 0 \\ \{b^\dagger(p,s), d^\dagger(p',s')\} = \{d(p,s), b(p',s')\} = 0\end{gathered} \tag{2.21}$$

and where

$$\sum_{\pm s} u_\alpha(p, s)\, \bar{u}_\beta(p, s) = (2m)^{-1}(\not{p} + m)_{\alpha\beta} \tag{2.22}$$

$$\sum_{\pm s} v_\alpha(p, s)\, \bar{v}_\beta(p, s) = (2m)^{-1}(\not{p} - m)_{\alpha\beta}$$

Eqs. 2.17 – 2.22 implement the canonical anticommutation relations

$$\{\psi^\dagger_a(x), \psi_b(y)\} = \delta_{ab}\,\delta^3(x - y) \tag{2.23}$$

$$\{\psi_a(x), \psi_b(y)\} = \{\psi^\dagger_a(x), \psi^\dagger_b(y)\} = 0$$

Generalization to Tachyon Solution

In this section we will define spin ½ tachyon fields quantized on a spacelike surface. We will see that localization problems and a non-canonical equal time anticommutator appear as noted in previous attempts. In the next chapter we will develop a light-front quantization program that enables us to define a local quantum field theory of tachyons without the problems of spacelike surface quantization.

We now define tachyon fields using the tachyon spinors defined in section 2.2:

$$\psi_T(x) = \sum_{\pm s} \int d^3p\ N_T(p)[b_T(p, s)u_T(p, s)e^{-ip.x} + d_T^\dagger(p, s)v_T(p, s)e^{ip.x}] \tag{2.24}$$

$$\psi_T^\dagger(x) = \sum_{\pm s} \int d^3p\ N_T(p)[b_T^\dagger(p, s)u_T^\dagger(p,s)e^{+ip.x} + d_T(p, s)v_T^\dagger(p, s)e^{-ip.x}] \tag{2.25}$$

where the subscript T designates "Tachyon", where the momenta satisfy

$$p_\mu p^\mu = p^{0\,2} - p^{x\,2} - p^{y\,2} - p^{z\,2} = -m^2 \tag{2.26}$$

and where

$$N_T(p) = [m/((2\pi)^3|\mathbf{p}|)]^{½} \tag{2.27}$$

The non-zero canonical equal-time anticommutation relation implied by the Lagrangian formulation is

$$\{\psi_{T\ a}^{\ \dagger}(x), \psi_{Tb}(x')\} = -[\gamma^5]_{ab}\, \delta^3(x - x') \tag{1.120}$$

We will now evaluate the equal-time commutation relations following from eqs. 2.24 and 2.25.

$$\{\psi_{T\ a}^{\ \dagger}(x), \psi_{Tb}(y)\} = \sum_{\pm s, s'} \int d^3p' d^3p\ N_T(p)N_T(p')\cdot$$

$$\cdot[\{b_T^\dagger(p',s'),b_T(p,s)\}u_{Ta}(p,s)u_{T\ b}^{\ \dagger}(p',s')e^{+ip.y - ip'.x} +$$

$$+ \{d_T(p',s'),d_T^\dagger(p,s)\}v_{Ta}(p,s)v_{T\ b}^{\ \dagger}(p',s')e^{-ip.y + ip'.x}]$$

$$= \sum_{\pm s} \int d^3p\ N_T^2(p)[u_{Ta}(p,s)u_{T\ b}^{\ \dagger}(p,s)e^{+ip.(y - x)} + v_{Ta}(p,s)v_{T\ b}^{\ \dagger}(p,s)e^{-ip.(y - x)}]$$

$$= i\int d^3p\ N_T^2(p)(2m|\mathbf{p}|)^{-1}[(-i\not{p} + m)\boldsymbol{\gamma}\cdot\mathbf{p}e^{-ip.(y - x)} + (-i\not{p} - m)\boldsymbol{\gamma}\cdot\mathbf{p}e^{+ip.(y - x)}]$$

$$= \int d^3p\ N_T^2(p)(|\mathbf{p}|/m + i\boldsymbol{\gamma}\cdot\mathbf{p}/|\mathbf{p}|)e^{-ip.(y - x)}$$

after letting $\mathbf{p} \rightarrow -\mathbf{p}$. Then

$$= \int d^3p\ \theta(\mathbf{p}^2 - m^2)(1 + im\boldsymbol{\gamma}\cdot\mathbf{p}/|\mathbf{p}|^2)e^{-ip.(y - x)}/(2\pi)^3$$

$$= \delta_{ab}\delta_T^{\ 3}(x - y) + im\int d^3p\ \theta(\mathbf{p}^2 - m^2)[\boldsymbol{\gamma}\cdot\mathbf{p}]_{ab}e^{-ip\cdot(y-x)}/[(2\pi)^3|\mathbf{p}|^2] \tag{2.28}$$

where

$$\delta_T^{\ 3}(x - y) = \int d^3p\ \theta(\mathbf{p}^2 - m^2)e^{ip\cdot(x-y)} \tag{2.29}$$

similar to Feinberg's[111] results. The restriction on the momentum integration does not allow a canonical $\delta^3(x - y)$, and leads to localization and other issues in this spacelike surface quantization. We show these issues are eliminated by using other coordinates to quantize on a non-spacelike surface called a *light-front*.

The second term in eq. 2.28 can be eliminated *for left-handed and right-handed* tachyon field commutators:

$$\{\psi_{TL}^{\ \dagger}(x), \psi_{TL}(y)\} = ½(1 - \gamma^5)\delta_T^{\ 3}(x - y) \tag{2.30}$$
$$\{\psi_{TR}^{\ \dagger}(x), \psi_{TR}(y)\} = ½(1 + \gamma^5)\delta_T^{\ 3}(x - y) \tag{2.31}$$

using the definitions of eq. 1.112. However, we still have

$$\{\psi_{TL}^{\ \dagger}(x), \psi_{TR}(y)\} = ½im\,(1 - \gamma^5)\int d^3p\ \theta(\mathbf{p}^2 - m^2)\boldsymbol{\gamma}\cdot\mathbf{p}e^{-ip\cdot(y-x)}/[(2\pi)^3|\mathbf{p}|^2] \tag{2.32}$$

Eqs. 2.30 and 2.31, together with the canonical anticommutation relation eq. 1.120, suggest that tachyon quantization should be done with left-handed and right-handed fields.

Therefore we will break off discussion of tachyon quantization on spacelike surfaces at this point, and discuss left-handed and right-handed tachyon light-front quantization in the next chapter.

---

[111] G. Feinberg, Phys. Rev. **159**, 1089 (1967.

3. Light-Front, "Handed" Tachyon Theory

There have been many studies of light-front (infinite momentum frame) physics in the past forty years.[112] In this chapter we show that the spin ½ tachyon formulation in the preceding chapter if separated into left-handed and right-handed fields, and light-front quantized has canonical anticommutation relations and an acceptable local, quantum field theoretic formulation. This development enables us to establish a framework for the derivation of the overall form of ElectroWeak theory in the next chapter.

In this chapter, in addition to spin ½ tachyons, we will discuss spin 0 and spin 1 tachyons. We will also consider discrete symmetries C, P, and T; causality; locality; the nature of particle states; the handling of negative energy states; Lorentz covariance; and perturbation theory.

3.1 Tachyon Quantization on the Light-Front

Light-front coordinates cannot be obtained by a Lorentz transformation, or by a Superluminal transformation, from a standard set of coordinate system variables even in a limiting sense. Instead they are a defined set of variables that have been used to develop quantum field theories that have been shown to be equivalent to quantum field theories based on conventional coordinates. In particular, light-front quantum field theories have been shown to yield fully Lorentz covariant S matrix elements that are the same as S matrix elements calculated in the conventional way.

Light-front variables are defined by:

$$x^{\pm} = (x^0 \pm x^3)/\surd 2 \qquad (3.1)$$
$$\partial/\partial x^{\pm} \equiv \partial^{\mp} \equiv (\partial/\partial x^0 \pm \partial/\partial x^3)/\surd 2$$

with the "transverse" coordinate variables, x^1 and x^2, unchanged.

Quantization on surfaces of constant x^+ (light-front surfaces) has been shown to support satisfactory formulations of Quantum Electrodynamics and other quantum field theories. Thus x^+ plays the role of the "time" variable in light-front quantized theories. We will define canonical equal x^+ anticommutation relations for fermions and spin ½ tachyons.

The inner product of two 4-vectors has the form

$$x \cdot y = x^+y^- + y^+x^- - x^1y^1 - x^2y^2 \qquad (3.2)$$

[112] L. Susskind, Phys. Rev. **165**, 1535 (1968); K. Bardakci and M. B. Halpern Phys. Rev. **176**, 1686 (1968), S. Weinberg, Phys. Rev. **150**, 1313 (1966); J. Kogut and D. Soper, Phys. Rev. **D1**, 2901 (1970); J. D. Bjorken, J. Kogut, and D. Soper, Phys. Rev. **D3**, 1382 (1971); R. A. Neville and F. Rohrlich, Nuov. Cim. **A1**, 625 (1971); F. Rohrlich, Acta Phys Austr. Suppl. **8**, 277 (1971); S-J Chang, R. Root, and T-M Yan, Phys. Rev. **D7**, 1133 (1973); S-J Chang, and T-M Yan, Phys. Rev. **D7**, 1147 (1973); T-M Yan, Phys. Rev. **D7**, 1761 (1973); T-M Yan, Phys. Rev. **D7**, 1780 (1973); C. Thorn, Phys. Rev. **D19**, 639 (1979); and references therein.

and the light-front definition of Dirac matrices (eqs. 1.56) is analogous to eq. 3.1:

$$\gamma^{\pm} = (\gamma^0 \pm \gamma^3)/\sqrt{2} \tag{3.3}$$

$$\gamma^{\pm\,2} = 0$$

and with transverse matrices γ^1 and γ^2 defined as usual.

Light-Front Dirac Fermion Second Quantization

In this subsection we will outline the well-known, conventional light-front canonical quantization of spin ½ Dirac particles. Our purpose will be to set the stage for the light-front canonical quantization of spin ½ tachyons in the following section.

We begin with the Lagrangian density

$$\mathcal{L} = \bar{\psi}(i\gamma^{\mu}\partial/\partial x^{\mu} - m)\psi(x) \tag{3.4}$$

and action

$$L = \int d^4x\mathcal{L} \equiv \int dx^{+}dx^{-}d^2x\mathcal{L} \tag{3.5}$$

written in terms of light-front variables. Variation of the action yields the free Dirac equation:

$$(i\gamma^{+}\partial/\partial x^{-} + i\gamma^{-}\partial/\partial x^{+} + i\gamma^{j}\partial/\partial x^{j} - m)\psi(x) = 0 \tag{3.6}$$

This equation can be decomposed into equations, which explicitly show that only two components of the Dirac wave function are independent. The relevant projection operators are:

$$\begin{aligned}
&R^{\pm} = \tfrac{1}{2}(I \pm \gamma^0\gamma^3) \\
&R^{+} + R^{-} = I \\
&R^{\pm\,2} = R^{\pm} \\
&R^{\pm\,\dagger} = R^{\pm} \\
&R^{+}R^{-} = 0 \\
&\gamma^{\pm} = \sqrt{2}\,\gamma^0R^{\pm} \\
&\gamma^{\pm}\gamma^{\mp} = 2R^{\mp}
\end{aligned} \tag{3.7}$$

When they are applied to the Dirac wave function they yield two wave functions:

$$\psi^{\pm}(x) = R^{\pm}\psi(x) \tag{3.8}$$

even in the presence of interactions (as shown by Yan et al and Kogut et al). The free field equations of $\psi^{\pm}$ that are implied by eq. 3.6, after applying the projection operators, are

$$i\partial^{-}\psi^{+} - 2^{-1/2}(i\gamma^{j}\partial/\partial x^{j} + m)\gamma^{0}\psi^{-} = 0 \quad (3.9)$$

$$i\partial^{+}\psi^{-} - 2^{-1/2}(i\gamma^{j}\partial/\partial x^{j} + m)\gamma^{0}\psi^{+} = 0 \quad (3.10)$$

where the factor of $2^{1/2}$ is due to the factor of $2^{-1/2}$ in the defintion of $\partial^{\pm}$. Eq. 3.10 can be integrated to express ψ^{-} in terms of ψ^{+}

$$\psi^{-}(x) = i \int dy^{-}d^{2}y\ \varepsilon(x^{-} - y^{-})\delta^{2}(x - y)(i\gamma^{j}\partial/\partial y^{j} + m)\gamma^{0}\psi^{+}(y)/2^{5/2} \quad (3.11)$$

and thus ψ^{-} and $\psi^{-\dagger}$ are dependent fields. Consequently the canonical equal light-front ($x^{+} = y^{+}$) anticommutation relations of the independent fields are:

$$\{\psi^{+}_{a}(x), \psi^{+\dagger}_{b}(y)\} = 2^{-1/2}R^{+}_{ab}\,\delta(x^{-} - y^{-})\delta^{2}(x - y) \quad (3.12)$$

$$\{\psi^{+}_{a}(x), \psi^{+}_{b}(y)\} = \{\psi^{+\dagger}_{a}(x), \psi^{+\dagger}_{b}(y)\} = 0$$

where the factor of $2^{-1/2}$ is a result of the definition of x^{-} in eq. 3.1 and the delta function identity

$$\delta(x^{-}) = 2^{1/2}\delta(x^{0} - x^{3}) \quad (3.13)$$

The light-front form of the free Lagrangian is

$$\begin{aligned}\mathfrak{L} = {} & \psi^{+\dagger}i\partial^{-}\psi^{+} - 2^{-1/2}\psi^{+\dagger}(i\gamma^{j}\partial_{j} + m)\gamma^{0}\psi^{-} + \\ & + \psi^{-\dagger}i\partial^{+}\psi^{-} - 2^{-1/2}\psi^{-\dagger}(i\gamma^{j}\partial_{j} + m)\gamma^{0}\psi^{+}\end{aligned} \quad (3.14)$$

and the conjugate momenta are

$$\pi^{+} = \partial\mathfrak{L}/\partial(\partial^{-}\psi^{+}) = i\psi^{+\dagger} \quad (3.15)$$

$$\pi^{-} = \partial\mathfrak{L}/\partial(\partial^{-}\psi^{-}) = 0 \quad (3.16)$$

which is consistent with the canonical equal-light-front ($x^{+} = y^{+}$) anticommutation relations (eq. 3.12).

The Hamiltonian density is

$$\mathcal{H} = \pi^{+}\partial^{-}\psi^{+} - \mathfrak{L} \quad (3.17)$$

The Fourier representation of the independent field and its hermitean conjugate are

$$\overset{+}{\psi}(x) = \sum_{\pm s} \int d^2p dp^+ \overset{+}{N}(p)\theta(p^+)[\overset{+}{b}(p, s)\overset{+}{u}(p, s)e^{-ip.x} + \overset{+}{d}{}^{\dagger}(p, s)\overset{+}{v}(p, s)e^{ip.x}] \tag{3.18}$$

$$\overset{+}{\psi}{}^{\dagger}(x) = \sum_{\pm s} \int d^2p dp^+ \overset{+}{N}(p)\theta(p^+)[\overset{+}{b}{}^{\dagger}(p, s)\overset{+}{u}{}^{\dagger}(p,s)e^{+ip.x} + \overset{+}{d}(p, s)\overset{+}{v}{}^{\dagger}(p, s)e^{-ip.x}] \tag{3.19}$$

where

$$\overset{+}{N}(p) = [m/((2\pi)^3 p^+)]^{1/2} \tag{3.20}$$

and where the anticommutation relations of the Fourier coefficient operators are

$$\begin{aligned}
\{\overset{+}{b}(p,s), \overset{+}{b}{}^{\dagger}(p',s')\} &= \delta_{ss'}\delta^2(\mathbf{p} - \mathbf{p'})\delta(p^+ - p'^+) \\
\{\overset{+}{d}(p,s), \overset{+}{d}{}^{\dagger}(p',s')\} &= \delta_{ss'}\delta^2(\mathbf{p} - \mathbf{p'})\delta(p^+ - p'^+) \\
\{\overset{+}{b}(p,s), \overset{+}{b}(p',s')\} &= \{\overset{+}{d}(p,s), \overset{+}{d}(p',s')\} = 0 \\
\{\overset{+}{b}{}^{\dagger}(p,s), \overset{+}{b}{}^{\dagger}(p',s')\} &= \{\overset{+}{d}{}^{\dagger}(p,s), \overset{+}{d}{}^{\dagger}(p',s')\} = 0 \\
\{\overset{+}{b}(p,s), \overset{+}{d}{}^{\dagger}(p',s')\} &= \{\overset{+}{d}(p,s), \overset{+}{b}{}^{\dagger}(p',s')\} = 0 \\
\{\overset{+}{b}{}^{\dagger}(p,s), \overset{+}{d}{}^{\dagger}(p',s')\} &= \{\overset{+}{d}(p,s), \overset{+}{b}(p',s')\} = 0
\end{aligned} \tag{3.21}$$

and where

$$\overset{+}{u}(p, s) = \overset{+}{R}u(p,s) \tag{3.22}$$

$$\overset{+}{v}(p, s) = \overset{+}{R}v(p,s) \tag{3.23}$$

with normalizations

$$\begin{aligned}
\overset{+}{u}{}^{\dagger}(p, s')\overset{+}{u}(p, s) &= u^{\dagger}(p, s')\overset{+}{R}{}^{2}u(p, s) = u^{\dagger}(p, s')\overset{+}{R}u(p, s) \\
&= 2^{-1/2}\bar{u}(p, s')\gamma^{+}u(p, s) = 2^{-1/2}\delta_{ss'}p^{+}/m
\end{aligned} \tag{3.24}$$

$$\overset{+}{v}{}^{\dagger}(p, s')\overset{+}{v}(p, s) = 2^{-1/2}\delta_{ss'}p^{+}/m \tag{3.25}$$

and

$$\sum_{\pm s} \overset{+}{u}_{\alpha}(p, s)\, \overset{-+}{u}_{\beta}(p, s) = [\overset{+}{R}(2m)^{-1}(\not{p} + m)\overset{-}{R}]_{\alpha\beta} \tag{3.26}$$

$$\sum_{\pm s} \overset{+}{v}_{\alpha}(p, s)\, \overset{-+}{v}_{\beta}(p, s) = [\overset{+}{R}(2m)^{-1}(\not{p} - m)\overset{-}{R}]_{\alpha\beta} \tag{3.27}$$

where $\overset{-+}{u}(p, s) = u^{\dagger}(p, s)\overset{+}{R}\gamma^0$. The evaluation of the equal light-front anticommutator yields eq. 3.12.

The light-front Feynman propagator is defined to be

$$iS_{LFF}(x,y)\gamma^0 = \theta(x^+ - y^+)<0|\psi(x)\psi^\dagger(y)|0> - \theta(y^+ - x^+)<0|\psi^\dagger(y)\psi(x)|0> \quad (3.28)$$

with spinor indices understood. After some calculation we find

$$\begin{aligned} iS_{LFF}(x, y) &= iS_F(x, y) - 2^{-5/2}\gamma^+\in(x^- - y^-)\delta(x^+ - y^+)\delta^2(x - y) \\ &= iS_F(x, y) - \gamma^0R^+\in(x^- - y^-)\delta(x^+ - y^+)\delta^2(x - y)/4 \end{aligned} \quad (3.29)$$

where $iS_F(x, y)$ is the conventional fermion Feynman propagator.

If interactions are introduced, light-front perturbation theory has been shown to lead to the same results as conventional perturbation theory with the effects of the second term in eq. 3.29 cancelled in perturbation theory.[113]

Light-Front Spin ½ Tachyon Second Quantization

In this subsection we will resume the discussion of spin ½ tachyons and consider light-front tachyon quantization. We will see that the incompleteness of the set of tachyon wave functions within the framework of space-like coordinate quantization is not present in the case of light-front quantization. Thus we will be able to implement canonical commutation relations and have localized tachyons. We will follow the same line of light-front development for tachyons as we did for Dirac fermions in the preceding subsection so that the reader can compare the development step by step.

In chapter 2 we developed a spin ½ tachyon field Fourier expansion:

$$\psi_T(x) = \sum_{\pm s} \int d^3p\, N_T(p)[b_T(p, s)u_T(p, s)e^{-ip.x} + d_T^\dagger(p, s)v_T(p, s)e^{+ip.x}] \quad (2.24)$$

where

$$N_T(p) = [m/((2\pi)^3|\mathbf{p}|)]^{1/2} \quad (2.27)$$

We now change to light-front variables for the primed coordinates (eq. 3.1) as in the preceding subsection.

Since the tachyon wave functions satisfy the tachyon Dirac equation eq. 1.89:

$$(\gamma^\mu\partial/\partial x^\mu + m)\psi_T(x) = 0 \quad (1.89)$$

[113] S-J Chang, R. Root, and T-M Yan, Phys. Rev. **D7**, 1133 (1973); S-J Chang, and T-M Yan, Phys. Rev. **D7**, 1147 (1973); T-M Yan, Phys. Rev. **D7**, 1761 (1973); T-M Yan, Phys. Rev. **D7**, 1780 (1973); J. Kogut and D. Soper, Phys. Rev. **D1**, 2901 (1970); J. D. Bjorken, J. Kogut, and D. Soper, Phys. Rev. **D3**, 1382 (1971); C. Thorn, Phys. Rev. **D19**, 639 (1979); and references therein.

we can simply determine the tachyon version of the light-front Dirac equations (eqs. 3.9 and 3.10) by omitting the factors of i and changing the sign of m:

$$\partial^-\psi_T{}^+ - 2^{-1/2}(\gamma^j\partial/\partial x^j - m)\gamma^0\psi_T{}^- = 0 \tag{3.30}$$

$$\partial^+\psi_T{}^- - 2^{-1/2}(\gamma^j\partial/\partial x^j - m)\gamma^0\psi_T{}^+ = 0 \tag{3.31}$$

where

$$\psi_T{}^\pm(x) = R^\pm\psi_T(x) \tag{3.32}$$

The independent field is $\psi_T{}^+$ (as in the Dirac field case), and thus the *canonical* equal-light-front ($x^+ = y^+$) anticommutation relations for the independent field are:

$$\{\psi_T{}^+{}_a(x), \psi_T{}^{+\dagger}{}_b(y)\} = 2^{-1/2}R^+{}_{ab}\,\delta(x^- - y^-)\delta^2(x - y) \tag{3.33}$$

$$\{\psi_T{}^+{}_{ba}(x), \psi_T{}^+{}_b(y)\} = \{\psi_T{}^{+\dagger}{}_a(x), \psi_T{}^{+\dagger}{}_b(y)\} = 0 \tag{3.34}$$

Starting from the tachyon wave function Fourier expansion (eq. 2.24) we define the light-front Fourier expansion to be:

$$\psi_T{}^+(x) = \sum_{\pm s}\int d^2p\,dp^+N_T{}^+(p)\theta(p^+)[b_T{}^+(p, s)u_T{}^+(p, s)e^{-ip\cdot x} + + d_T{}^{+\dagger}(p, s)v_T{}^+(p, s)e^{+ip\cdot x}] \tag{3.35}$$

$$\psi_T{}^{+\dagger}(x) = \sum_{\pm s}\int d^2p\,dp^+N_T{}^+(p)\theta(p^+)\,[b_T{}^{+\dagger}(p, s)u_T{}^{+\dagger}(p, s)e^{+ip\cdot x} + + d_T{}^+(p, s)v_T{}^{+\dagger}(p, s)e^{-ip\cdot x}] \tag{3.36}$$

where

$$N_T{}^+(p) = [2m|\mathbf{p}|/((2\pi)^3(p^+(p^+ - p^-) + p_\perp{}^2))]^{1/2} \tag{3.37}$$

where the anticommutation relations of the Fourier coefficient operators are

$$\begin{aligned}
&\{b_T{}^+(q, s), b_T{}^{+\dagger}(p', s')\} = \delta_{ss'}\delta^2(\mathbf{q} - \mathbf{p}')\delta(q^+ - p'^+)\\
&\{d_T{}^+(q, s), d_T{}^{+\dagger}(p', s')\} = \delta_{ss'}\delta^2(\mathbf{q} - \mathbf{p}')\delta(q^+ - p'^+)\\
&\{b_T{}^+(q, s), b_T{}^+(p', s')\} = \{d_T{}^+(q, s), d_T{}^+(p', s')\} = 0\\
&\{b_T{}^{+\dagger}(q, s), b_T{}^{+\dagger}(p', s')\} = \{d_T{}^{+\dagger}(q, s), d_T{}^{+\dagger}(p', s')\} = 0\\
&\{b_T{}^+(q, s), d_T{}^{+\dagger}(p', s')\} = \{d_T{}^+(q, s), b_T{}^{+\dagger}(p', s')\} = 0\\
&\{b_T{}^{+\dagger}(q, s), d_T{}^{+\dagger}(p', s')\} = \{d_T{}^+(q, s), b_T{}^+(p', s')\} = 0
\end{aligned} \tag{3.38}$$

and where the spinors are

$$u_T^{\ +}(p, s) = R^{+}u_T(p, s) \tag{3.39}$$
$$v_T^{\ +}(p, s) = R^{+}v_T(p, s) \tag{3.40}$$
$$u_T^{\ +\dagger}(p, s) = u_T^{\ \dagger}(p, s)R^{+} \tag{3.41}$$
$$v_T^{\ +\dagger}(p, s) = v_T^{\ \dagger}(p, s)R^{+} \tag{3.42}$$

The evaluation of the anticommutator of eqs. 3.35 and 3.36 is

$$\{\psi_T{}^{+}{}_{a}(x), \psi_T{}^{+\dagger}{}_{b}(y)\} = \sum_{\pm s,s'} \int d^2p dp^{+} \int d^2p' dp'^{+}\ N_T{}^{+}(p)N_T{}^{+}(p')\theta(p^{+})\theta(p'^{+})\cdot$$
$$\cdot[\{b_T{}^{+\dagger}(p',s'),b_T{}^{+}(p,s)\}u_T{}^{+}{}_{a}(p,s)u_T{}^{+\dagger}{}_{b}(p',s')e^{+ip'\cdot y - ip\cdot x} +$$
$$+ \{d_T{}^{+}(p',s'),d_T{}^{+\dagger}(p,s)\}v_T{}^{+}{}_{a}(p,s)v_T{}^{+\dagger}{}_{b}(p',s')e^{-ip'\cdot y + ip\cdot x}]$$

$$= \sum_{\pm s} \int d^2p dp^{+} \int d^2p' dp'^{+}\ N_T{}^{+}(p)N_T{}^{+}(p')\theta(p^{+})\theta(p'^{+})\cdot$$
$$\cdot\delta^2(\mathbf{p}' - \mathbf{p})\delta(p'^{+} - p^{+})[u_T{}^{+}{}_{a}(p,s)u_T{}^{+\dagger}{}_{b}(p',s)e^{+ip'\cdot y - ip\cdot x} +$$
$$+ v_T{}^{+}{}_{a}(p,s)v_T{}^{+\dagger}{}_{b}(p',s)e^{-ip'\cdot y + ip\cdot x}]$$

$$= \sum_{\pm s} \int d^2p dp^{+}\ N_T{}^{+2}(p)\theta(p^{+})[u_T{}^{+}{}_{a}(p,s)u_T{}^{+\dagger}{}_{b}(p,s)e^{+ip\cdot(y-x)} +$$
$$+ v_T{}^{+}{}_{a}(p,s)v_T{}^{+\dagger}{}_{b}(p,s)e^{-ip\cdot(y-x)}]$$

$$= i\int d^2p dp^{+}\theta(p^{+})N_T{}^{+2}(p)(2m|\mathbf{p}|)^{-1}\{[R^{+}(-i\not{p} + m)\boldsymbol{\gamma}\cdot\mathbf{p}R^{+}]_{ab}e^{+ip\cdot(y-x)} +$$
$$+ [R^{+}(-i\not{p} - m)\boldsymbol{\gamma}\cdot\mathbf{p}R^{+}]_{ab}e^{-ip\cdot(y-x)}\}$$

$$= i\int d^2p_{\perp}\int_0^{\infty} dp^{+}N_T{}^{+2}(p)\{[R^{+}(-ip^{+}(p^{+} - p^{-}) - ip_{\perp}{}^2 + mp_{\perp}\cdot\gamma_{\perp})]_{ab}e^{+ip^{+}(y^{-} - x^{-}) - ip_{\perp}\cdot(y_{\perp} - x_{\perp})}$$
$$- [R^{+}(ip^{+}(p^{+} - p^{-}) + ip_{\perp}{}^2 + mp_{\perp}\cdot\gamma_{\perp})]_{ab}e^{-ip^{+}(y^{-} - x^{-}) + ip_{\perp}\cdot(y_{\perp} - x_{\perp})}\}/(2m|\mathbf{p}|)$$

$$= \int d^2p_\perp \int_{-\infty}^{\infty} dp^+ N_T^{+2}(p)[R^+(p^+(p^+ - p^-) + p_\perp{}^2 + imp_\perp \cdot \gamma_\perp)]_{ab} \cdot \cdot e^{+ip^+(y^- - x^-) - ip_\perp \cdot (y_\perp - x_\perp)}/(2m|\mathbf{p}|)$$

upon letting $p^+ \to -p^+$ and $\mathbf{p}_\perp \to -\mathbf{p}_\perp$ in the second term. Continuing we find it equals

$$= \int d^2p_\perp \int_{-\infty}^{\infty} dp^+ (2\pi)^{-3}[R^+(1 + imp_\perp \cdot \gamma_\perp/(p^+(p^+ - p^-) + p_\perp{}^2))]_{ab} e^{+ip^+(y^- - x^-) - ip_\perp \cdot (y_\perp - x_\perp)}$$

$$= R^+{}_{ab}\delta(y^- - x^-)\delta^2(\mathbf{y} - \mathbf{x}) + \delta_T{}^+(y - x) \tag{3.43}$$

where $\mathbf{p}_\perp' = (p_1, p_2)$ and where

$$\delta_T{}^+(y - x) = im \int d^2p_\perp \int_{-\infty}^{\infty} dp^+[R^+p_\perp \cdot \gamma_\perp]_{ab} e^{+ip^+(y^- - x^-) - ip_\perp \cdot (y_\perp - x_\perp)}/[(2\pi)^3(p^+(p^+ - p^-) + p_\perp{}^2)] \tag{3.44}$$

Note that the use of light-front variables enables us to obtain the δ-functions implied by the canonical quantization procedure. As a result we *almost* obtain eqs. 3.33 and 3.34 as the equal-light-front ($x^+ = y^+$) anticommutation relations for the independent field. However there is an extra term $\delta_T{}^+(y - x)$ that is not canonical and, more importantly, these anticommutation relations are not appropriate for tachyons because the correct tachyon Lagrangian does not lead to eqs. 3.33 and 3.34. The correct tachyon Lagrangian requires a canonical conjugate momentum that is given by eq. 1.95 and is not $i\psi_T{}^{+\dagger}(x')$ as eq. 3.33 indirectly implies. The presence of a γ^5 factor suggests that we must separate the light-front tachyons into left-handed and right-handed parts. The solution of these issues requires light-front quantized, left-handed and right-handed tachyon fields!

3.2 Light-Front, Left-Handed & Right-Handed Tachyon Quantization

Again, the spin ½ tachyon Dirac equation is

$$(\gamma^\mu \partial/\partial x^\mu + m)\psi_T(x) = 0 \tag{1.89}$$

It can be derived using the canonical procedure from the Lagrangian, and action,

$$\mathcal{L} = \psi_T{}^S(\gamma^\mu \partial/\partial x^\mu + m)\psi_T(x) \tag{1.91}$$

$$I = \int d^4x \mathcal{L}_T \tag{1.92}$$

where

$$\psi_T{}^S = \psi_T{}^\dagger i\gamma^0\gamma^5 \tag{1.93}$$

We will now take the Lagrangian density of eq. 1.91, and decompose it into left-handed and right-handed parts with a view towards obtaining canonical commutation relations. We will use the projection operators:

$$\begin{aligned} &C^\pm = \tfrac{1}{2}(I \pm \gamma^5) \\ &C^+ + C^- = I \\ &C^{\pm\,2} = C^\pm \\ &C^+C^- = 0 \\ &\psi_{TL} = C^-\psi_T \\ &\psi_{TR} = C^+\psi_T \end{aligned} \tag{3.45}$$

After decomposing the Lagrangian we will rewrite it in terms of light-front variables. The decomposition into left-handed and right-handed wave fields using $C^\pm$ is:

$$\mathcal{L}_T = \psi_{TL}{}^\dagger\gamma^0 i\gamma^\mu\partial_\mu\psi_{TL} - \psi_{TR}{}^\dagger\gamma^0 i\gamma^\mu\partial_\mu\psi_{TR} - im[\psi_{TR}{}^\dagger\gamma^0\psi_{TL} - \psi_{TL}{}^\dagger\gamma^0\psi_{TR}] \tag{3.46}$$

Now if we transform to light-front variables we obtain the light-front tachyon Lagrangian:

$$\begin{aligned} \mathcal{L}_T = {}& 2^{1/2}\psi_{TL}{}^{+\dagger}i\partial^-\psi_{TL}{}^+ + 2^{1/2}\psi_{TL}{}^{-\dagger}i\partial^+\psi_{TL}{}^- - \psi_{TL}{}^{+\dagger}\gamma^0 i\gamma^j\partial^j\psi_{TL}{}^- - \psi_{TL}{}^{-\dagger}\gamma^0 i\gamma^j\partial^j\psi_{TL}{}^+ - \\ & - 2^{1/2}\psi_{TR}{}^{+\dagger}i\partial^-\psi_{TR}{}^+ - 2^{1/2}\psi_{TR}{}^{-\dagger}i\partial^+\psi_{TR}{}^- + \psi_{TR}{}^{+\dagger}\gamma^0 i\gamma^j\partial^j\psi_{TR}{}^- + \psi_{TR}{}^{-\dagger}\gamma^0 i\gamma^j\partial^j\psi_{TR}{}^+ - \\ & - im[\psi_{TR}{}^{+\dagger}\gamma^0\psi_{TL}{}^- - \psi_{TL}{}^{+\dagger}\gamma^0\psi_{TR}{}^- + \psi_{TR}{}^{-\dagger}\gamma^0\psi_{TL}{}^+ - \psi_{TL}{}^{-\dagger}\gamma^0\psi_{TR}{}^+] \end{aligned} \tag{3.47}$$

with implied sums over j = 1,2. In contrast to the light-front tachyon Lagrangian we note the corresponding Dirac fermion Lagrangian is

$$\begin{aligned} \mathcal{L}_D = {}& 2^{1/2}\psi_L{}^{+\dagger}i\partial^-\psi_L{}^+ + 2^{1/2}\psi_L{}^{-\dagger}i\partial^+\psi_L{}^- - \psi_L{}^{+\dagger}\gamma^0 i\gamma^j\partial^j\psi_L{}^- - \psi_L{}^{-\dagger}\gamma^0 i\gamma^j\partial^j\psi_L{}^+ - \\ & + 2^{1/2}\psi_R{}^{+\dagger}i\partial^-\psi_R{}^+ + 2^{1/2}\psi_R{}^{-\dagger}i\partial^+\psi_R{}^- - \psi_R{}^{+\dagger}\gamma^0 i\gamma^j\partial^j\psi_R{}^- - \psi_R{}^{-\dagger}\gamma^0 i\gamma^j\partial^j\psi_R{}^+ - \\ & - im[\psi_R{}^{+\dagger}\gamma^0\psi_L{}^- + \psi_L{}^{+\dagger}\gamma^0\psi_R{}^- + \psi_R{}^{-\dagger}\gamma^0\psi_L{}^+ + \psi_L{}^{-\dagger}\gamma^0\psi_R{}^+] \end{aligned} \tag{3.48}$$

The difference in signs between eqs. 3.47 and 3.48 will turn out to be a crucial factor in the derivation of the form of ElectroWeak theory in the next chapter.

Returning to the tachyon Lagrangian eq. 3.47 we can obtain equations of motion through the standard variational techniques:

$$
\begin{aligned}
2^{\frac{1}{2}\dagger} i\partial^- \psi_{TL}{}^+ - \gamma^0 i\gamma^j \partial^j \psi_{TL}{}^- + im\gamma^0 \psi_{TR}{}^- &= 0 \\
2^{\frac{1}{2}} i\partial^- \psi_{TR}{}^+ - \gamma^0 i\gamma^j \partial^j \psi_{TR}{}^- + im\gamma^0 \psi_{TL}{}^- &= 0 \\
2^{\frac{1}{2}} i\partial^+ \psi_{TL}{}^- - \gamma^0 i\gamma^j \partial^j \psi_{TL}{}^+ + im\gamma^0 \psi_{TR}{}^+ &= 0 \\
2^{\frac{1}{2}} i\partial^+ \psi_{TR}{}^- - \gamma^0 i\gamma^j \partial^j \psi_{TR}{}^+ + im\gamma^0 \psi_{TL}{}^+ &= 0
\end{aligned}
\tag{3.49}
$$

Eqs. 3.49 show that $\psi_{TL}{}^-$ and $\psi_{TR}{}^-$ are dependent fields that are functions of $\psi_{TL}{}^+$ and $\psi_{TR}{}^+$ on the light-front where x^+ equals a constant. They can be expressed in an integral form similar to eq. 3.11. (The independent fields $\psi_{TL}{}^+$ and $\psi_{TR}{}^+$ play a fundamental role in tachyon theory and can be used to define "in" and "out" tachyon states in perturbation theory.)

The conjugate momenta implied by eq. 3.47 are

$$
\begin{aligned}
\pi_{TL}{}^+ &= \partial\mathfrak{L}/\partial(\partial^- \psi_{TL}{}^+) = 2^{\frac{1}{2}} i\psi_{TL}{}^{+\dagger} \qquad (3.50) \\
\pi_{TL}{}^- &= \partial\mathfrak{L}/\partial(\partial^- \psi_{TL}{}^-) = 0 \\
\pi_{TR}{}^+ &= \partial\mathfrak{L}/\partial(\partial^- \psi_{TR}{}^+) = -2^{\frac{1}{2}} i\psi_{TR}{}^{+\dagger} \qquad (3.51) \\
\pi_{TR}{}^- &= \partial\mathfrak{L}/\partial(\partial^- \psi_{TR}{}^-) = 0
\end{aligned}
$$

The resulting canonical equal-light-front ($x^+ = y^+$) anticommutation relations of the independent fields are:

$$
\{\psi_{TL}{}^{+\dagger}{}_a(x), \psi_{TL}{}^+{}_b(y)\} = 2^{-1}[C^- R^+]_{ab}\, \delta(x^- - y^-)\delta^2(x - y) \tag{3.52}
$$

$$
\{\psi_{TR}{}^{+\dagger}{}_a(x), \psi_{TR}{}^+{}_b(y)\} = -2^{-1}[C^+ R^+]_{ab}\, \delta(x^- - y^-)\delta^2(x - y) \tag{3.53}
$$

$$
\{\psi_{TL}{}^{+\dagger}{}_a(x), \psi_{TR}{}^+{}_b(y)\} = \{\psi_{TR}{}^{+\dagger}{}_a(x), \psi_{TL}{}^+{}_b(y)\} = 0 \tag{3.54}
$$

$$
\{\psi_{TL}{}^+{}_a(x), \psi_{TR}{}^+{}_b(y)\} = \{\psi_{TR}{}^{+\dagger}{}_a(x), \psi_{TL}{}^{+\dagger}{}_b(y)\} = 0 \tag{3.55}
$$

where the factors of 2^{-1} are the result of the $2^{\frac{1}{2}}$ factor in eqs. 3.50 and 3.51, and the factor of $2^{-\frac{1}{2}}$ in the definition of x^- in eq. 3.1.

If we compare eqs. 3.52 and 3.53 with the corresponding anticommutation relations of conventional Dirac quantum fields

$$\{\psi_L{}^{+\dagger}{}_a(x), \psi_L{}^{+}{}_b(y)\} = 2^{-1}[C^-R^+]_{ab}\,\delta(x^- - y^-)\delta^2(x - y) \tag{3.52a}$$

$$\{\psi_R{}^{+\dagger}{}_a(x), \psi_R{}^{+}{}_b(y)\} = 2^{-1}[C^+R^+]_{ab}\,\delta(x^- - y^-)\delta^2(x - y) \tag{3.53a}$$

we see that the right-handed tachyon anticommutation relation (eq. 3.53) has a minus sign relative to the right-handed conventional anticommutation relation (eq. 3.53a). The right-handed tachyon anticommutation relation (eq. 3.53) with its minus sign will require compensating minus signs in its creation and annihilation Fourier component operators' commutation relations.

The sign differences between the Lagrangian terms in eqs. 3.47 and 3.48 will ultimately lead to a parity violating form for the ElectroWeak Lagrangian and thus resolve the long-standing question: Why parity violation? (See the following chapter.)

Left-Handed Tachyons

The left-handed tachyon wave function light-front Fourier expansion is:

$$\psi_{TL}{}^{+}(x) = \sum_{\pm s}\int d^2p dp^+ N_{TL}{}^{+}(p)\theta(p^+)[b_{TL}{}^{+}(p, s)u_{TL}{}^{+}(p, s)e^{-ip.x} + + d_{TL}{}^{+\dagger}(p, s)v_{TL}{}^{+}(p, s)e^{+ip.x}] \tag{3.56}$$

and its hermitean conjugate is

$$\psi_{TL}{}^{+\dagger}(x) = \sum_{\pm s}\int d^2p dp^+ N_{TL}{}^{+}(p)\theta(p^+)\,[b_{TL}{}^{+\dagger}(p, s)u_{TL}{}^{+\dagger}(p,s)e^{+ip.x} + + d_{TL}{}^{+}(p, s)v_{TL}{}^{+\dagger}(p, s)e^{-ip.x}] \tag{3.57}$$

where $N^+(p)$ is the same as $N_T{}^+(p)$ in eq. 3.37:

$$N_{TL}{}^{+}(p) = [m|\mathbf{p}|/((2\pi)^3(p^+(p^+ - p^-) + p_\perp{}^2))]^{1/2} \tag{3.57a}$$

where the anticommutation relations of the Fourier coefficient operators are (The factors of $2^{-1/2}$ appearing in eqs. 3.58 are due to eq. 3.13.)

$$\begin{aligned}
&\{b_{TL}{}^{+}(q,s), b_{TL}{}^{+\dagger}(p,s')\} = 2^{-1/2}\delta_{ss'}\delta^2(\mathbf{q} - \mathbf{p})\delta(q^+ - p^+)\\
&\{d_{TL}{}^{+}(q,s), d_{TL}{}^{+\dagger}(p,s')\} = 2^{-1/2}\delta_{ss'}\delta^2(\mathbf{q} - \mathbf{p})\delta(q^+ - p^+)\\
&\{b_{TL}{}^{+}(q,s), b_{TL}{}^{+}(p,s')\} = \{d_{TL}{}^{+}(q,s), d_{TL}{}^{+}(p,s')\} = 0\\
&\{b_{TL}{}^{+\dagger}(q,s), b_{TL}{}^{+\dagger}(p,s')\} = \{d_{TL}{}^{+\dagger}(q,s), d_{TL}{}^{+\dagger}(p,s')\} = 0\\
&\{b_{TL}{}^{+}(q,s), d_{TL}{}^{+\dagger}(p,s')\} = \{d_{TL}{}^{+}(q,s), b_{TL}{}^{+\dagger}(p,s')\} = 0\\
&\{b_{TL}{}^{+\dagger}(q,s), d_{TL}{}^{+\dagger}(p,s')\} = \{d_{TL}{}^{+}(q,s), b_{TL}{}^{+}(p,s')\} = 0
\end{aligned} \tag{3.58}$$

and where the spinors are

$$\begin{aligned} u_{TL}^{+}(p, s) &= C^{-} u_{T}^{+}(p,s) \\ v_{TL}^{+}(p, s) &= C^{-} v_{T}^{+}(p,s) \\ u_{TL}^{+\dagger}(p, s) &= u_{T}^{+\dagger}(p, s)C^{-} \\ v_{TL}^{+\dagger}(p, s) &= v_{T}^{+\dagger}(p, s)C^{-} \end{aligned} \tag{3.59}$$

using eqs. 3.39 through 3.42. The canonical left-handed anticommutation relation (eq. 3.52) follows from eqs. 3.56 and 3.57.

$$\{\psi_{TL}{}^{+}{}_{a}(x), \psi_{TL}{}^{+\dagger}{}_{b}(y)\} = \sum_{\pm s,s'} \int d^2p dp^{+} \int d^2p' dp'^{+}\, N_{TL}{}^{+}(p) N_{TL}{}^{+}(p')\theta(p^{+})\theta(p'^{+})\cdot$$

$$\cdot[\{b_{TL}{}^{+\dagger}(p',s'), b_{TL}{}^{+}(p,s)\}u_{TL}{}^{+}{}_{a}(p,s)u_{TL}{}^{+\dagger}{}_{b}(p',s')e^{+ip'\cdot y - ip\cdot x} +$$

$$+ \{d_{TL}{}^{+}(p',s'), d_{TL}{}^{+\dagger}(p,s)\}v_{TL}{}^{+}{}_{a}(p,s)v_{TL}{}^{+\dagger}{}_{b}(p',s')e^{-ip'\cdot y + ip\cdot x}]$$

$$= \sum_{\pm s} \int d^2p dp^{+}\, N_{TL}{}^{+2}(p)\theta(p^{+})[u_{TL}{}^{+}{}_{a}(p,s)u_{TL}{}^{+\dagger}{}_{b}(p,s)e^{+ip\cdot(y-x)} +$$

$$+ v_{TL}{}^{+}{}_{a}(p,s)v_{TL}{}^{+\dagger}{}_{b}(p,s)e^{-ip\cdot(y-x)}]$$

$$= i\int d^2p dp^{+}\theta(p^{+})N_{TL}{}^{+2}(p)(2m|\mathbf{p}|)^{-1}\{[C^{-}R^{+}(-i\not{p} + m)\boldsymbol{\gamma}\cdot\mathbf{p}R^{+}C^{-}]_{ab}e^{+ip\cdot(y-x)} +$$

$$+ [C^{-}R^{+}(-i\not{p} - m)\boldsymbol{\gamma}\cdot\mathbf{p}R^{+}C^{-}]_{ab}e^{-ip\cdot(y-x)}\}$$

$$= i\int d^2p_{\perp}\int_0^{\infty} dp^{+} N_{TL}{}^{+2}(p)\{[C^{-}R^{+}(-ip^{+}(p^{+} - p^{-}) - ip_{\perp}{}^{2} + mp_{\perp}\cdot\gamma_{\perp})C^{-}]_{ab}\cdot$$

$$\cdot e^{+ip^{+}(y^{-} - x^{-}) - ip_{\perp}\cdot(y_{\perp} - x_{\perp})} -$$

$$- [C^{-}R^{+}(ip^{+}(p^{+} - p^{-}) + ip_{\perp}{}^{2} + mp_{\perp}\cdot\gamma_{\perp})C^{-}]_{ab}e^{-ip^{+}(y^{-} - x^{-}) + ip_{\perp}\cdot(y_{\perp} - x_{\perp})}\}/(2m|\mathbf{p}|)$$

$$= \int d^2p_{\perp}\int_{-\infty}^{\infty} dp^{+} N_{TL}{}^{+2}(p)[C^{-}R^{+}(p^{+}(p^{+} - p^{-}) + p_{\perp}{}^{2})]_{ab}\cdot$$

$$\cdot e^{+ip^{+}(y^{-} - x^{-}) - ip_{\perp}\cdot(y_{\perp} - x_{\perp})}/(2m|\mathbf{p}|)$$

upon letting $p^+ \to -p^+$ and $\mathbf{p}_\perp \to -\mathbf{p}_\perp$ in the second term.

$$= \tfrac{1}{2}\int d^2p_\perp \int_{-\infty}^{\infty} dp^+ (2\pi)^{-3}[C^-R^+]_{ab} e^{+ip^+(y^- - x^-) - ip_\perp\cdot(y_\perp - x_\perp)}$$

$$= 2^{-1}[C^-R^+]_{ab}\,\delta(y^- - x^-)\delta^2(\mathbf{y} - \mathbf{x}) \tag{3.60}$$

Therefore we have left-handed, light-front quantized tachyons with canonical commutation relations and localized tachyons. A comparison with eq. 3.12 shows a clear similarity to Dirac fermion light-front second quantization.

Right-Handed Tachyons

The case of right-handed tachyons is quite similar to the left-handed case with only two differences: a minus sign in the creation and annihilation operator anticommutation relations, and right-handed projection operators. The right-handed tachyon wave function light-front Fourier expansion is:

$$\psi_{TR}^{+}(x) = \sum_{\pm s} \int d^2p dp^+ N_{TR}^{+}(p)\theta(p^+)[b_{TR}^{+}(p, s)u_{TR}^{+}(p, s)e^{-ip.x} + \\ + d_{TR}^{+\dagger}(p, s)v_{TR}^{+}(p, s)e^{+ip.x}] \tag{3.61}$$

and its hermitean conjugate is

$$\psi_{TR}^{+\dagger}(x) = \sum_{\pm s} \int d^2p dp^+ N_{TR}^{+}(p)\theta(p^+) [b_{TR}^{+\dagger}(p, s)u_{TR}^{+\dagger}(p, s)e^{+ip.x} + \\ + d_{TR}^{+}(p, s)v_{TR}^{+\dagger}(p, s)e^{-ip.x}] \tag{3.62}$$

where $N_{TR}^{+}(p) = N_{TL}^{+}(p)$, where the anticommutation relations of the Fourier coefficient operators are (The factors of $2^{-1/2}$ below are due to eq. 3.13.)

$$\{b_{TR}^{+}(q,s), b_{TR}^{+\dagger}(p,s')\} = -2^{-1/2}\delta_{ss'}\delta^2(\mathbf{q} - \mathbf{p})\delta(q^+ - p^+) \tag{3.63}$$
$$\{d_{TR}^{+}(q,s), d_{TR}^{+\dagger}(p,s')\} = -2^{-1/2}\delta_{ss'}\delta^2(\mathbf{q} - \mathbf{p})\delta(q^+ - p^+)$$
$$\{b_{TR}^{+}(q,s), b_{TR}^{+}(p,s')\} = \{d_{TR}^{+}(q,s), d_{TR}^{+}(p,s')\} = 0$$
$$\{b_{TR}^{+\dagger}(q,s), b_{TR}^{+\dagger}(p,s')\} = \{d_{TR}^{+\dagger}(q,s), d_{TR}^{+\dagger}(p,s')\} = 0$$
$$\{b_{TR}^{+}(q,s), d_{TR}^{+\dagger}(p,s')\} = \{d_{TR}^{+}(q,s), b_{TR}^{+\dagger}(p,s')\} = 0$$
$$\{b_{TR}^{+\dagger}(q,s), d_{TR}^{+\dagger}(p,s')\} = \{d_{TR}^{+}(q,s), b_{TR}^{+}(p,s')\} = 0$$

and where the spinors are

$$u_{TR}^{+}(p, s) = C^{+}u_{T}^{+}(p,s) \tag{3.64}$$

$$v_{TR}^{+}(p, s) = C^{+}v_{T}^{+}(p,s) \tag{3.65}$$

using eqs. 3.43 and 3.44 with normalizations given by eqs. 3.24 – 3.27.

The right-handed anticommutation relation (eq. 3.53) with the minus sign follows in particular because of the minus signs in eqs. 3.63.

Interpretation of Tachyon Creation and Annihilation Operators

To properly discuss the physical interpretation of tachyon creation and annihilation operators we must first determine the Hamiltonian and momentum operators in terms of creation and annihilation operators.

The energy-momentum tensor density is the symmetrized version of

$$\mathfrak{T}^{\mu\nu} = \sum_i \partial\mathfrak{L}/\partial(\partial\chi_i/\partial x_\mu)\, \partial\chi_i/\partial x_\nu - g^{\mu\nu}\mathfrak{L} \tag{3.66}$$

where the sum over i is over the fields. The light-front Hamiltonian is

$$H \equiv P^- = T^{+-} = \int dx^- d^2x \mathfrak{T}^{+-} \tag{3.67}$$

And the "momenta" are

$$P^+ = T^{++} = \int dx^- d^2x \mathfrak{T}^{++} \tag{3.68}$$

$$P^i = T^{+i} = \int dx^- d^2x \mathfrak{T}^{+i} \tag{3.69}$$

for i = 1,2.

The light-front, left-handed and right-handed tachyon Lagrangian $\mathfrak{L}_T$ is eq. 3.47 and its equations of motion are eqs. 3.49. They imply

$$H = i2^{-1/2}\int dx^- d^2x\, [\psi_{TL}^{+\dagger}\partial^-\psi_{TL}^{+} - \partial^-\psi_{TL}^{+\dagger}\psi_{TL}^{+} + \psi_{TL}^{-\dagger}\partial^+\psi_{TL}^{-} - \partial^+\psi_{TL}^{-\dagger}\psi_{TL}^{-} -$$

$$- \psi_{TR}^{+\dagger}\partial^-\psi_{TR}^{+} + \partial^-\psi_{TR}^{+\dagger}\psi_{TR}^{+} - \psi_{TR}^{-\dagger}\partial^+\psi_{TR}^{-} + \partial^+\psi_{TR}^{-\dagger}\psi_{TR}^{-}] \tag{3.70}$$

After substituting for the various fields we find the *independent fields* (which constitute the in and out particle states) have the Hamiltonian terms:

$$H = \sum_{\pm s} \int d^2p dp^+\, p^-[b_{TL}^{+\dagger}(p,s)b_{TL}^{+}(p,s) - d_{TL}^{+}(p,s)d_{TL}^{+\dagger}(p,s) -$$

$$- b_{TR}^{+\dagger}(p,s)b_{TR}^{+}(p,s) + d_{TR}^{+}(p,s)d_{TR}^{+\dagger}(p,s)] \qquad (3.71)$$

$$= \sum_{\pm s}\int d^2p\,dp^+\, p^-[b_{TL}^{+\dagger}(p,s)b_{TL}^{+}(p,s) + d_{TL}^{+\dagger}(p,s)d_{TL}^{+}(p,s) -$$

$$- b_{TR}^{+\dagger}(p,s)b_{TR}^{+}(p,s) - d_{TR}^{+\dagger}(p,s)d_{TR}^{+}(p,s)] \qquad (3.72)$$

up to the usual infinite constants due to left-handed operator rearrangement and right-handed operator rearrangement that are discarded. Eq. 3.72 is the basis for our particle interpretation of tachyon creation and annihilation operators based on Dirac's hole theory. Dirac hole theory as applied in the light frame assumes all negative p^- ("energy") states are filled.

Left-Handed Tachyon Creation and Annihilation Operators

1. We identify $b_{TL}^{+\dagger}(p',s')$ and $d_{TL}^{+}(p',s')$ as creation operators for left-handed tachyons. $b_{TL}^{+\dagger}(p',s')$ creates a positive p^- ("energy") state and $d_{TL}^{+}(p',s')$ creates a negative p^- ("energy") state.

2. $b_{TL}^{+}(p',s')$ and $d_{TL}^{+\dagger}(p',s')$ are the corresponding annihilation operators for left-handed tachyons. $b_{TL}^{+}(p',s')$ annihilates a positive p^- ("energy") state and $d_{TL}^{+\dagger}(p',s')$ annihilates a negative p^- ("energy") state.

3. We assume Dirac hole theory holds for the left-handed tachyon vacuum with all negative energy states filled. There is no tachyon energy gap as there is for Dirac fermions. There is also the problem that the left-handed tachyon vacuum is not invariant under ordinary Lorentz transformations or Superluminal transformations. However if we confine ourselves to the infinite momentum frame for computations no ambiguity can result and the Lorentz covariant quantities that we calculate, such as the S matrix, are well-defined.

4. Using tachyon hole theory we identify $b_{TL}^{+}(p',s')$ and $d_{TL}^{+\dagger}(p',s')$ as annihilation operators for left-handed tachyons. $b_{TL}^{+}(p',s')$ annihilates a positive p^- ("energy") state and $d_{TL}^{+\dagger}(p',s')$ annihilates a negative p^- ("energy") state – thus creating a hole in the tachyon sea that we view as the creation of a positive p^- ("energy"), left-handed antitachyon. $d_{TL}^{+}(p',s')$ annihilates a positive p^- ("energy"), left-handed antitachyon.

Right-Handed Tachyons

The anticommutation relations of right-handed tachyon creation and annhilation operators (eqs. 3.63) and the right-handed Hamiltonian terms have the "wrong" sign compared to corresponding Dirac operators and left-handed tachyon operators. This situation is completely analogous to the situation of time-like photons in the covariant formulation of quantum Electrodynamics.[114] In the case of time-like photons it was possible to introduce an indefinite metric (Gupta-Bleuler formulation), and then to use the subsidiary condition $\partial A^{\nu}/\partial x^{\nu} = 0$ to reduce the dynamics of QED to the transverse components. Thus the time-like photons were intermediate artifacts needed to have a manifestly covariant formulation while QED observables depended solely on the transverse components of the electromagnetic field.

In the present case of free tachyons, and in leptonic ElectroWeak Theory there is no evident "subsidiary condition" to eliminate the right-handed tachyon fields. But since the only manner in which the right-handed leptonic tachyon fields[115] interact is through mass terms, which can be easily 'integrated out", right-handed leptonic tachyon fields are removed from the observable part of the leptonic ElectroWeak Theory by their "lack of interaction" with left-handed fields. In the case of quark ElectroWeak Theory right-handed tachyon quark fields have charge (–1/3) and thus experience an electromagnetic interaction as well as a Z interaction. However, since quarks are totally confined, right-handed tachyon quarks will not be able to continuously emit photons or Z's due to energy conservation and their confinement to bound states of fixed positive energy levels. Thus right-handed tachyons are analogous to time-like photons in the combined leptonic and quark ElectroWeak Theory – necessary theoretically but prevented from causing a negative energy disaster by the forms of their interactions. We discuss this subject in more detail in the following chapters.

3.3 Tachyon Feynman Propagator

In this section we develop the light-front propagator for tachyons. We begin with a subsection describing the light-front propagators of Dirac fields.

Dirac Field Light-Front Propagators

The light-front Feynman propagator for the complete field ψ of a Dirac fermion is given by eqs. 3.28 and 3.29. Eq. 3.29 shows that the propagator contains a non-covariant piece.

The light-front Feynman propagator for the ψ^+ field of a Dirac fermion is

$$iS^+_F(x,y)\gamma^0 = \theta(x^+ - y^+)\langle 0|\psi^+(x)\psi^{+\dagger}(y)|0\rangle - \theta(y^+ - x^+)\langle 0|\psi^{+\dagger}(y)\psi^+(x)|0\rangle \tag{3.73}$$

and does not contain a non-covariant piece due to the projection operators:

$$iS^+_F(x,y) = \int d^2p dp^+\theta(p^+)[1/(2(2\pi)^3p^+)]\{\theta(x^+ - y^+)[R^+(\not{p} + m)R^-]\, e^{-ip\cdot(x-y)} +$$
$$+ \theta(y^+ - x^+)[R^+(-\not{p} + m)R^-]e^{+ip\cdot(x-y)}\}$$

---

[114] Bogoliubov (1959) pp. 130-136.

[115] The tachyon fields are assumed to be neutrino fields in the leptonic sector, and d, s and b quarks in the quark sector.

$$= R^{+} iS_F(x,y)R^{-} \qquad (3.74)$$

where $S_F(x,y)$ is the usual Feynman propagator.

The light-front Feynman propagator for a *left-handed* Dirac field ψ^+ is

$$iS^{+}_{LF}(x,y) = \int d^2p dp^+ \theta(p^+)[1/(2(2\pi)^3 p^+)]\{\theta(x^+ - y^+)[C^- R^+(\not{p} + m)R^- C^-]e^{-ip\cdot(x-y)} + \\ + \theta(y^+ - x^+)[C^- R^+(-\not{p} + m)R^- C^-]e^{+ip\cdot(x-y)}\}$$

$$= C^- R^{+} iS_F(x,y)R^- C^- \qquad (3.75)$$

Tachyon Field Feynman Propagator

Turning now to tachyons, the light-front Feynman propagator for the left-handed $\psi_{TL}^{\;+}$ *tachyon* field is (using the Fourier expansion of the left-handed tachyon field eqs. 3.56 and 3.57):

$$iS^{+}_{TLF}(x,y) = \theta(x^+ - y^+)\langle 0|\psi_{TL}^{\;+}(x)\psi_{TL}^{\;+\dagger}(y)\gamma^0|0\rangle - \\ - \theta(y^+ - x^+)\langle 0|\psi_{TL}^{\;+\dagger}(y)\gamma^0\psi_{TL}^{\;+}(x)|0\rangle$$

$$= i\int d^2p dp^+ \theta(p^+) N_{TL}^{+2}(2m|\mathbf{p}|)^{-1} C^- R^+\{\theta(x^+ - y^+)[(-i\not{p} + m)\boldsymbol{\gamma}\cdot\mathbf{p}]e^{-ip\cdot(x-y)} + \\ + \theta(y^+ - x^+)[(-i\not{p} - m)\boldsymbol{\gamma}\cdot\mathbf{p}]e^{+ip\cdot(x-y)}\}R^+ C^- \gamma^0$$

If we define the on-shell momentum variable $p_0^- = (p_0^1 p_0^1 + p_0^2 p_0^2 - m^2)/(2p_0^+)$, $p_0^+ = p^+$, $p_0^j = p^j$ (for j = 1, 2), $p_{\perp 0}^{\;2} = p_0^j p_0^j$ and $\not{p}_0 = p_0\cdot\gamma$ then the above equation can be rewritten as

$$= iC^- R^+ \int d^4p[32\pi^4(p_0^+(p_0^+ - p_0^-) + p_{0\perp}^{\;2})]^{-1} e^{-ip\cdot(x-y)}\;\cdot$$

$$\cdot\{\theta(p^+)(-i\not{p}_0 + m)\boldsymbol{\gamma}\cdot\mathbf{p}_0]/[p^+(p^- - p_0^- + i\varepsilon)] +$$

$$+ \theta(-p^+)(-i\not{p}_0 + m)\boldsymbol{\gamma}\cdot\mathbf{p}_0]/[p^+(p^- + p_0^- - i\varepsilon)]\}R^+ C^- \gamma^0$$

$$= i\int d^4p(2\pi)^{-4}[C^- R^+(-i\not{p} + m)\boldsymbol{\gamma}\cdot\mathbf{p}R^+ C^- \gamma^0]e^{-ip\cdot(x-y)}\cdot \\ \cdot[(p^2 + m^2 + i\varepsilon)(2p^+(p^+ - p^-) + p_\perp^{\;2}))]^{-1}$$

$$= \tfrac{1}{2} C^- R^+ \gamma^0 \int d^4p(2\pi)^{-4} e^{-ip\cdot(x-y)}/(p^2 + m^2 + i\varepsilon)$$

$$= -2^{-3/2}C^{-}R^{+}\gamma^{0}\Delta_{FT}(x-y) \tag{3.76}$$

where

$$\Delta_{FT}(x-y) = -2^{\frac{1}{2}}\int d^4p(2\pi)^{-4}e^{-ip.(x-y)}/(p^2 + m^2 + i\varepsilon) \tag{3.77}$$

is the tachyon Feynman propagator for the "tachyon Klein-Gordon equation" (eq. 3.89 below).

Similarly the light-front Feynman propagator for the right-handed ψ_{TR}^{+} tachyon field is

$$iS^{+}_{TRF}(x,y) = \theta(x^{+} - y^{+})\langle 0|\psi_{TR}^{+}(x)\psi_{TR}^{+\dagger}(y)\gamma^{0}|0\rangle -$$
$$- \theta(y^{+} - x^{+})\langle 0|\psi_{TR}^{+\dagger}(y)\gamma^{0}\psi_{TR}^{+}(x)|0\rangle$$

$$= 2^{-3/2}C^{+}R^{+}\gamma^{0}\Delta_{FT}(x-y) \tag{3.78}$$

where the relative minus sign between eqs. 3.78 and 3.76 is due to the relative minus signs of the Fouier component operator anticommutation relations in eq. 3.58 and 3.63.

Thus we find *tachyon* pole terms in the tachyon propagators as one would expect.

3.4 Massive Scalar Tachyons

The case of massive scalar tachyons would normally be the starting point for the discussion of tachyon particles. But the possible connection of spin ½ tachyons to ElectroWeak Theory that was previewed in chapter 1 led us to consider spin ½ tachyons first. We now turn to free, neutral, spin 0 tachyons, which we anticipate would satisfy the tachyon equivalent of the Klein-Gordon equation:

$$(\Box - m^2)\phi_T(x) = 0 \tag{3.79}$$

where

$$\Box = \partial/\partial x_{\mu}\partial/\partial x^{\mu} \tag{3.80}$$

It can be derived using the canonical procedure from the Lagrangian density and action

$$\mathcal{L}_T = \tfrac{1}{2}[\partial\phi_T/\partial x^{\mu}\partial\phi_T/\partial x_{\mu} + m^2\phi_T^{\,2}] \tag{3.81}$$
$$I = \int d^4x\mathcal{L}_T$$

We will now proceed to canonically second quantize this theory using light-front coordinates (eq. 3.1). The Lagrangian density then becomes

$$\mathcal{L}_T = \partial\phi_T/\partial x^+ \partial\phi_T/\partial x^- - \tfrac{1}{2}\partial\phi_T/\partial x^i \partial\phi_T/\partial x^i + \tfrac{1}{2}m^2\phi_T{}^2 \tag{3.82}$$

The conjugate momentum is

$$\pi_T = \partial\mathcal{L}/\partial(\partial^-\phi_T) = \partial^+\phi_T \equiv \partial\phi_T/\partial x^- \tag{3.83}$$

and the equal x^+ commutation relations[116] are

$$[\pi_T(x), \phi_T(y)] = -i2^{-1/2}\delta(x^- - y^-)\delta^2(\mathbf{x}_\perp - \mathbf{y}_\perp) \tag{3.84}$$

We provisionally define the Fourier expansion of ϕ_T as

$$\phi_T(x) = \int d^2p dp^+ N_T(p)\theta(p^+)[a_T(p)e^{-ip.x} + a_T{}^\dagger(p)e^{+ip.x}] \tag{3.85}$$

where $N_T(p)$ is

$$N_T(p) = [(2\pi)^3 p^+]^{-1/2} \tag{3.86}$$

and the Fourier component operator commutation relations are

$$[a_T(q), a_T{}^\dagger(p)] = 2^{-1/2}\delta^2(\mathbf{q} - \mathbf{p'})\delta(q^+ - p^+) \tag{3.87}$$
$$[a_T(q), a_T(p)] = [a_T(q), a_T(p)] = 0$$

We now calculate

$$[\pi_T(x), \phi_T(y)] = [\partial\phi_T(x)/\partial x^-, \phi_T(y)]$$

$$= \int d^2p dp^+ \int d^2p' dp'^+ N_T(p)N_T(p')\theta(p^+)\theta(p'^+)\cdot$$

---

[116] Feinberg (G. Feinberg, Phys. Rev. **159**, 1089 (1967)) and others have suggested that scalar tachyons obey anticommutation relations because a Lorentz transformation can change a positive energy to a negative energy (and vice versa). However in light-front coordinates a Lorentz or Superluminal boost in the z direction does not change the sign of the equivalent of energy p^-. Boosts in other directions may change the sign of p^-. However the light-front is a particular choice of variables in a specific frame. Since perturbative and other calculations lead to covariant results we can do all calculations in one frame, and then, after expressing the results in covariant form, transform to any other reference frame. Then tachyon scattering events seen in the new coordinate system should be in agreement with the corresponding covariant calculation of the event. Therefore scalar tachyon quantization using commutators is acceptable and has the advantage of conforming to the general quantization program for scalar particles.

$$\cdot\{-ip^+[a_T(p), a_T^\dagger(p')]e^{+ip'.y - ip.x} + ip^+[a_T^\dagger(p), a_T(p')]e^{-ip'.y + ip.x}\}$$

$$= -i2^{-1/2}\int d^2p_\perp \int_0^\infty dp^+ N_T^{+2}(p)p^+\{e^{+ip^+(y^- - x^-) - ip_\perp\cdot(y_\perp - x_\perp)} + e^{-ip^+(y^- - x^-) + ip_\perp\cdot(y_\perp - x_\perp)}\}$$

$$= -i2^{-1/2}\int d^2p_\perp \int_{-\infty}^\infty dp^+(2\pi)^{-3}\, e^{+ip^+(y^- - x^-) - \mathbf{ip}_\perp\cdot(\mathbf{y}_\perp - \mathbf{x}_\perp)}$$

$$= -i2^{-1/2}\delta(x^- - y^-)\delta^2(\mathbf{x}_\perp - \mathbf{y}_\perp) \tag{3.88}$$

verifying the equal x^+ commutation relation.

Scalar Tachyon Feynman Propagator

The scalar tachyon Feynman propagator is defined by

$$i\Delta_{TF}(x - y) = \theta(x^+ - y^+)\langle 0|\phi_T(x)\, \phi_T(y)|0\rangle + \theta(y^+ - x^+)\langle 0|\phi_T(y)\phi_T(x)|0\rangle$$

$$= \int d^2p dp^+ \int d^2p'dp'^+\, N_T(p)N_T(p')\theta(p^+)\theta(p'^+)\cdot$$

$$\cdot\{\langle 0|a_T(p)a_T^\dagger(p')|0\rangle e^{+ip'.y - ip.x} + \langle 0|a_T(p')a_T^\dagger(p)|0\rangle e^{-ip'.y + ip.x}\}$$

$$= 2^{-1/2}\int d^2p_\perp \int_0^\infty dp^+ N_T^{+2}(p)\{e^{+ip^+(y^- - x^-) - ip_\perp\cdot(y_\perp - x_\perp)} + e^{-ip^+(y^- - x^-) + ip_\perp\cdot(y_\perp - x_\perp)}\}$$

$$= 2^{-1/2}\int d^2p_\perp \int_{-\infty}^\infty dp^+(2\pi)^{-3}\, e^{+ip^+(y^- - x^-) - \mathbf{ip}_\perp\cdot(\mathbf{y}_\perp - \mathbf{x}_\perp)}/p^+$$

$$= -i2^{1/2}\int d^4p(2\pi)^{-4}e^{-ip\cdot(x - y)}/(p^2 + m^2 + i\varepsilon) = i\Delta_{FT}(x - y) \tag{3.89}$$

in agreement with the definition in eq. 3.77.

3.5 Massive Vector Tachyons

The case of massive vector tachyons is of some interest since massive vector bosons, W and Z bosons, have been found in nature. Therefore there is a possibility that, hitherto undiscovered, massive vector tachyons may exist in nature and might eventually be created by particle accelerators. In this section we will second quantize a massive tachyon vector boson in light-front coordinates.

We begin with a standard, neutral, free, massive vector boson Lagrangian with the sign of the mass term changed to make it a tachyon vector boson Lagrangian:

$$\mathcal{L}_{TVB} = -\tfrac{1}{4} F_T^{\mu\nu}(x)F_{T\mu\nu}(x) - \tfrac{1}{2} m^2 V_T^{\mu} V_{T\mu} \tag{3.90}$$

where

$$F_{T\mu\nu} = (\partial V_{T\mu}/\partial x^{\nu} - \partial V_{T\nu}/\partial x^{\mu}) \tag{3.91}$$

The equations of motion implied by eq. 3.90 are

$$\partial F_T^{\mu\nu}/\partial x^{\nu} - m^2 V_T^{\mu} = 0 \tag{3.92}$$

Eq. 3.92 implies the subsidiary condition

$$\partial V_T^{\mu}/\partial x^{\mu} = 0 \tag{3.93}$$

which, in turn, implies

$$(\Box - m^2)V_T^{\mu} = 0 \tag{3.94}$$

after substituting eq. 3.91 in eq. 3.90 and using eq. 3.93, where

$$\Box = \partial/\partial x_{\mu}\partial/\partial x^{\mu} \tag{3.95}$$

as previously.

Eq. 3.94 is immediately recognizable as a tachyon equation for each component. We now transform the Lagrangian to light-front coordinates and proceed to quantize. Using the previous definition of light-front variables (eq. 3.1) we define the fields:

$$\begin{aligned} F_T^{+-} &= \partial^{+}V_T^{-} - \partial^{-}V_T^{+} \\ F_T^{+j} &= \partial^{+}V_T^{j} - \partial^{j}V_T^{+} \\ F_T^{-j} &= \partial^{-}V_T^{j} - \partial^{j}V_T^{-} \\ F_T^{ij} &= \partial^{i}V_T^{j} - \partial^{j}V_T^{i} \\ V_T^{-} &= 2^{-1/2}(V_T^{0} - V_T^{3}) \\ V_T^{+} &= 2^{-1/2}(V_T^{0} + V_T^{3}) \end{aligned} \tag{3.96}$$

The light-front equivalent of the Lagrangian (eq. 3.90) is:

$$\mathcal{L}_{TVB} = -\tfrac{1}{2}(\partial V_{T\mu}/\partial x^{\nu}\partial V_T{}^{\mu}/\partial x_{\nu} - \partial V_T{}^{\mu}/\partial x_{\nu}\partial V_{T\nu}/\partial x^{\mu}) - \tfrac{1}{2}m^2 V_T{}^{\mu}V_{T\mu}$$

After using the constraint eq. 3.93 and discarding a total divergence, we see

$$\mathcal{L}_{TVB} \equiv -\tfrac{1}{2}\partial V_{T\mu}/\partial x^{\nu}\partial V_T{}^{\mu}/\partial x_{\nu} - \tfrac{1}{2}m^2 V_T{}^{\mu}V_{T\mu}$$

which becomes

$$\mathcal{L}_{TVB} \equiv -\partial^{+}V_T{}^{-}\partial^{-}V_T{}^{+} - \partial^{+}V_T{}^{+}\partial^{-}V_T{}^{-} + \partial^{+}V_T{}^{i}\partial^{-}V_T{}^{i} + \partial^{i}V_T{}^{+}\partial^{i}V_T{}^{-} + \\ + \tfrac{1}{2}\partial^{i}V_T{}^{j}\partial^{i}V_T{}^{j} - \tfrac{1}{2}m^2(2V_T{}^{+}V_T{}^{-} - V_T{}^{i}V_T{}^{i}) \tag{3.97}$$

using light-front coordinates with implied sums over i and j. The resulting equations of motion are

$$(\Box - m^2)V_T{}^{-} = 0 \tag{3.98}$$
$$(\Box - m^2)V_T{}^{+} = 0$$
$$(\Box - m^2)V_T{}^{i} = 0$$

for i = 1, 2 as one can also see directly from eq. 3.94.

The conjugate spacelike-surface momenta are

$$\pi^{\mu} = \partial\mathcal{L}/\partial(\partial^0 V_T{}^{\mu}) = -\,\partial V_T{}^{\mu}/\partial x^0 \tag{3.99}$$

and the conjugate light-front momenta are

$$\pi^{+} = \partial\mathcal{L}/\partial(\partial^{-}V_T{}^{+}) = -\,\partial^{+}V_T{}^{-} \equiv -\,\partial V_T{}^{-}/\partial x^{-} \tag{3.100}$$
$$\pi^{-} = \partial\mathcal{L}/\partial(\partial^{-}V_T{}^{-}) = -\,\partial^{+}V_T{}^{+} \equiv -\,\partial V_T{}^{+}/\partial x^{-} \tag{3.101}$$
$$\pi^{i} = \partial\mathcal{L}/\partial(\partial^{-}V_T{}^{i}) = \partial^{+}V_T{}^{i} \equiv \partial V_T{}^{i}/\partial x^{-} \tag{3.102}$$

The equal x^{+} commutation relations are

$$[\pi_T{}^{a}(x), V_T{}^{b}(y)] = -i2^{-\frac{1}{2}}\delta^{3ab}(x - y) \tag{3.103}$$

where

$$\delta^{3ab}(x - y) = \int d^2k dk^{+} e^{i[k+(x- - y-) - \mathbf{k}.(\mathbf{x} - \mathbf{y})]}[g^{ab} + k^a k^b/m^2]/(2\pi)^3 \tag{3.104}$$

where $\mathbf{k} = (k^1, k^2)$, and $g^{-+} = g^{+-} = 1 = -g^{11} = = -g^{22}$ with all other $g^{ab} = 0$. The equal x^+ commutation relations satisfy the constraint:

$$\partial^a[\pi_T^a(x), V_T^b(y)] = \partial^b[\pi_T^a(x), V_T^b(y)] = 0 \tag{3.105}$$

implied by eq. 3.93.

Next we define the Fourier expansion of V_T^μ as

$$V_T^\mu(x) = \sum_s \int d^2k dk^+ N_{TV}(k)\theta(k^+)\varepsilon^\mu(k, s)[a_T(k, s)e^{-ik\cdot x} + a_T^\dagger(k, s)e^{+ik\cdot x}] \tag{3.106}$$

where $k^2 = 2k^+k^- - k^{i\,2} = -m^2$, and where $N_{TV}(k)$ is

$$N_{TV}(k) = [(2\pi)^3 k^+]^{-½} \tag{3.107}$$

There are three spin orientations: two transverse orientations and a longitudinal orientation, $s = \pm1, 0$. The spin polarization vector satisfies

$$k_\mu\varepsilon^\mu(k, s) = 0 \tag{3.108}$$

It also satisfies the normalization condition

$$\sum_s \varepsilon^\mu(k, s)\varepsilon^\upsilon(k, s) = -(g^{\mu\nu} + k^\mu k^\nu/m^2) \tag{3.109}$$

The Fourier component operator commutation relations are

$$[a_T(q, s), a_T^\dagger(p, s')] = 2^{-½}\delta_{ss'}\delta^2(\mathbf{q} - \mathbf{p'})\delta(q^+ - p^+) \tag{3.110}$$
$$[a_T(q, s), a_T(p, s')] = [a_T(q, s), a_T(p, s')] = 0$$

Eqs. 3.106 – 3.110 imply the commutation relations eqs. 3.103.

Vector Tachyon Feynman Propagator

The vector tachyon Feynman propagator is defined by

$$i\Delta_{TF}(x - y)^{\mu\nu} = \theta(x^+ - y^+)\langle 0|V_T^\mu(x)V_T^\nu(y)|0\rangle +$$
$$+ \theta(y^+ - x^+)\langle 0|V_T^\nu(y)V_T^\mu(x)|0\rangle \tag{3.111}$$

and found to equal

$$= -i \int \frac{d^4k\, e^{-ik.(x-y)}\, (g^{\mu\nu} + k^\mu k^\nu/m^2)}{(2\pi)^4\, (k^2 + m^2 + i\varepsilon)}$$

The propagator displays the tachyon poles as expected.

3.6 Massive Spin 2 Tachyons – Massive Tachyon Gravitons

Gravitons – the quanta of gravitation – are massless as far as we know. Massive gravitons have been a subject of a number of theoretical investigations. While there is no evidence for massive gravitons there is evidence that the universe in the large has additional forces at play that affect the rotation of galaxies and seem to be producing an accelerating expansion of the universe. Therefore it is sensible to consider the possibility that massive spin 2 tachyons may exist that could play a role in the understanding of the unusual features of the universe in the large. Since the effect of new forces seems to be seen only at distances comparable to the size of galaxies or greater, it is possible that massive spin 2 tachyons may have a small mass of the order of [1/L] where L is the galactic radius of galaxies such as our galaxy. We leave the calculation of the propagator and other details as an exercise for the reader.

3.7 Tachyons and the Discrete Symmetries: C, P, and T

The discrete (improper) transformations, parity, time reversal and charge conjugation, are of major importance in analyzing the structure of ElectroWeak Theory and the Standard Model. In this section we will examine these transformations with respect to tachyon particles of various spins.

First, bosonic tachyons (spins 0, 1, and 2) have discrete transformation properties similar to ordinary bosons and so will not be considered further. The interested reader should read standard texts on this topic and notice the sign of the squared mass does not introduce any distinctive differences between tachyon and normal bosons.

In the case of fermions (odd half integer spin particles) there is a difference between tachyon fermions and conventional fermions. We will consider the case of spin ½ tachyons. Fundamental tachyon fermions of higher spin (should any exist) would also have distinctively different P, C, and T transformation properties.

We will use the manifestly covariant Lagrangian

$$\mathcal{L} = \psi_T^{\dagger} i\gamma^0\gamma^5(\gamma^\mu\partial/\partial x^\mu + m)\psi_T(x) \quad (1.91)$$

and equations of motion

$$(\gamma^\mu\partial/\partial x^\mu + m)\psi_T(x) = 0 \quad (1.89)$$

and anticommutator

$$\{\psi_{T\,a}^{\dagger}(x), \psi_{Tb}(x')\} = -[\gamma^5]_{ab}\,\delta^3(x - x') \qquad (1.120)$$

as the starting points of our discussion of discrete transformation properties.

Parity

In defining the parity transformation for spin ½ tachyons we try to retain as much similarity as possible to the Dirac spin ½ fermion parity transformation. By definition the parity transformation changes **x** → –**x**. In the case of a Dirac field if the transformation is defined as:

$$\mathcal{P}\psi(\mathbf{x}, t)\mathcal{P}^{-1} = \gamma^0\psi(-\mathbf{x}, t) \qquad (3.112)$$

then the free Dirac field Lagrangian, field equation and and anticommutators are invariant under this transformation.

If we now consider a spin ½ tachyon field and assume the same form for the transformation:

$$\mathcal{P}\psi_T(\mathbf{x}, t)\mathcal{P}^{-1} = \gamma^0\psi_T(-\mathbf{x}, t) \qquad (3.113)$$

then we find

$$\mathcal{P}\mathcal{L}(\mathbf{x}, t)\mathcal{P}^{-1} = -\mathcal{L}(-\mathbf{x}, t) \qquad (3.114)$$

$$(\gamma^\mu\partial/\partial x'^\mu + m)\psi_T(-\mathbf{x}, t) = 0 \qquad (3.115)$$

$$\{\psi_{T\,a}^{\dagger}(x'), \psi_{Tb}(y')\} = [\gamma^5]_{ab}\,\delta^3(x' - y') \qquad (1.120)$$

where x' = (–**x**, t) and y' = (–**y**, t). The Lagrangian and the anticommutation relations (compare to eq. 1.120) change sign under the parity transformation. Therefore the physics of tachyons is not invariant under parity. This fact is evident from eq. 3.47 where the expression of the Lagrangian in terms of left-handed and right-handed fields shows the Lagrangian changes sign under the interchange of left and right handed fields (an effect of the parity transformation). Thus spin ½ tachyon theory, like nature, violates parity.

Note that parity violation is intrinsic to tachyons – even free tachyons. The discussion of ElectroWeak Theory in the following chapters will associate tachyon parity violation with ElectroWeak parity violation.

Charge Conjugation

The charge conjugation transformation is connected to the interchange of particle and antiparticle. If we assume that a spin ½ tachyon has charge and is coupled to the electromagnetic field then (assuming the usual gauge coupling) the tachyon Lagrangian becomes

$$\mathcal{L} = \psi_T^{\dagger} i\gamma^0\gamma^5[\gamma^\mu(\partial/\partial x^\mu + ieA_\mu) + m]\psi_T(x) \tag{3.116}$$

If the theory were charge conjugation invariant then a unitary operator $\mathcal{C}$ would exist that would change the sign of the electromagnetic current $j^\mu(x)$, and the electromagnetic field A(x, t), while leaving the Lagrangian invariant:

$$\mathcal{C}j^\mu(x)\mathcal{C}^{-1} = -j^\mu(x) \qquad ???? \tag{3.117}$$

$$\mathcal{C}\mathbf{A}(\mathbf{x}, t)\mathcal{C}^{-1} = -\mathbf{A}(\mathbf{x}, t) \qquad ???? \tag{3.118}$$

$$\mathcal{C}\mathcal{L}(\mathbf{x}, t)\mathcal{C}^{-1} = \mathcal{L}(\mathbf{x}, t) \qquad ???? \tag{3.119}$$

The tachyon electromagnetic current implied by the Lagrangian eq. 3.116 is

$$j_T^\mu(x) = \tfrac{1}{2} e[\psi_T^{\dagger}\gamma^0\gamma^5, \gamma^\mu\psi_T] \tag{3.120}$$

where we antisymmetrize as in the case of Dirac fermions.

We extend the standard charge conjugation transformation[117] *with one modification* from that of a conventional spin ½ field to the case of spin ½ tachyons:

$$\psi_{TC}(\mathbf{x}, t) = \mathcal{C}\psi_T(\mathbf{x}, t)\mathcal{C}^{-1} = C_T\bar{\psi}_T^{\,T}(\mathbf{x}, t) = C_T\gamma^0\psi_T^{\,*}(\mathbf{x}, t) \tag{3.121}$$

where $\psi_{TC}(\mathbf{x}, t)$ is the antitachyon field of opposite charge, where the superscript T indicating the transpose, and where the tachyon charge conjugation matrix (which differs from the Dirac field analogue) is

$$C_T = i\gamma^2\gamma^5\gamma^0 \qquad \text{and} \qquad C_T^{-1} = i\gamma^0\gamma^5\gamma^2 = -C_T = C_T^{\dagger} \tag{3.122}$$

The complex conjugate of the antitachyon field is

$$\psi_{TC}^{\dagger}(\mathbf{x}, t) = \mathcal{C}_T\psi_T^{\dagger}(\mathbf{x}, t)\mathcal{C}_T^{-1} = \bar{\psi}_T^{\,*}(\mathbf{x}, t)C_T^{-1} = \psi_T^{\,T}(\mathbf{x}, t)\gamma^0 C_T^{-1} \tag{3.123}$$

Under this transformation we find the tachyon Lagrangian, field equation, and anticommutator are invariant under charge conjugation:

[117] in the Dirac representation. See Bjorken (1965) p. 115, or Kaku (1993) p. 117.

$$[\gamma^{\mu}(\partial/\partial x^{\mu} + ieA_{\mu}) + m]\psi_{TC}(x) = 0 \tag{3.124}$$

using $\mathcal{C}_T A_\mu \mathcal{C}_T^{-1} = -A_\mu$, and for the Lagrangian in eq. 3.116

$$\mathcal{C}_T \mathcal{L}(\mathbf{x}, t)\mathcal{C}_T^{-1} = \mathcal{L}(\mathbf{x}, t) \tag{3.125}$$

The charge conjugate anticommutator is

$$\{\psi_{TC}{}^{\dagger}{}_{a}(x), \psi_{TCb}(y)\} = [\gamma^5]_{ab}\, \delta^3(x - y) \tag{3.126}$$

The charge conjugated current

$$j_{TC}{}^{\mu}(x) = \tfrac{1}{2} e[\psi_{TC}{}^{\dagger}\gamma^0\gamma^5, \gamma^{\mu}\psi_{TC}] = -\, j_T{}^{\mu}(x) \tag{3.127}$$

so that

$$\mathcal{C}_T j_T{}^{\mu}(x)A_\mu \mathcal{C}_T^{-1} = j_T{}^{\mu}(x)A_\mu \tag{3.128}$$

resulting in the tachyon charge conjugation invariant Lagrangian eq. 3.116. Thus tachyons and antitachyons can be expected to have the same charge and mass.

CP Transformation

The CP transformation has been of major theoretical and experimental interest for some time. Experimentally CP violation has been found in certain sectors: K meson and B meson decays.

Since we have seen that tachyons violate parity and do not violate charge conjugation we can see that tachyons inherently violate CP invariance.

Time Reversal

Time reversal invariance is also a significant theoretical and experimental issue. The standard Dirac fermion time reversal transformation is:

$$\mathfrak{T}\psi(\mathbf{x}, t)\mathfrak{T}^{-1} = T\psi(\mathbf{x}, -t) \tag{3.129}$$

where $\mathfrak{T} = \mathcal{U}K$ where $\mathcal{U}$ is a unitary operator and K is the operator that takes the complex conjugate of all c-numbers, and where

$$T = i\gamma^1\gamma^3 \tag{3.130}$$

Due to the form of $\mathcal{L}$ (eq. 3.116), which assumes an electromagnetic interaction for the purpose of illustration, we find that the tachyon time reversal transformation is

$$\mathfrak{T}_T\psi_T(\mathbf{x}, t)\mathfrak{T}_T^{-1} = T_T\psi_T(\mathbf{x}, -t) \tag{3.131}$$

where $\mathfrak{T}_T = \mathcal{U}_T K$ where $\mathcal{U}_T$ is a unitary operator and K is the operator that takes the complex conjugate of all c-numbers, and where

$$T_T = i\gamma^5\gamma^1\gamma^3 \tag{3.132}$$

The matrix T_T satisfies:

$$T_T^{-1} = -i\gamma^3\gamma^1\gamma^5 = T_T \tag{3.133}$$

$$T_T\gamma^\mu T_T^{-1} = -\gamma_\mu \tag{3.134}$$

Under time reversal the current satisfies

$$\mathfrak{T}_T j_{T\mu}(\mathbf{x}, t)\mathfrak{T}_T^{-1} = j_T^{\ \mu}(\mathbf{x}, -t) \tag{3.135}$$

If we assume the electromagnetic field satisfies

$$\mathfrak{T}\mathbf{A}(\mathbf{x}, t)\mathfrak{T}^{-1} = -\mathbf{A}(\mathbf{x}, -t)$$

under time reversal, then the tachyon Lagrangian

$$\mathcal{L} = \psi_T^{\dagger} i\gamma^0\gamma^5[\gamma^\mu(\partial/\partial x^\mu + ieA_\mu) + m]\psi_T(x) \tag{3.116}$$

satisfies

$$\mathfrak{T}_T\mathcal{L}(\mathbf{x}, t)\mathfrak{T}_T^{-1} = \mathcal{L}(\mathbf{x}, -t) \tag{3.136}$$

under time reversal. Although the action changes by a translation in time, Poincaré translation invariance implies the action is invariant. The equation of motion derived from the Lagrangian eq. 3.116 is also invariant under the tachyon time reversal transformation.

Thus we find the dynamics of the tachyon Lagrangian theory (eq. 3.116) to be invariant under the tachyon time reversal transformation.

Tachyon CPT Invariance and Tachyon-Extended CPT Theorem

The question of CPT invariance has long been of theoretical and experimental interest. For conventional particle theories the CPT Theorem implies CPT invariance under very general conditions. We will examine the case of CPT invariance of a model tachyon theory with Lagrangian eq. 3.116. For Dirac fermions

$$\mathcal{C}\mathcal{P}\mathfrak{T}\psi_a(\mathbf{x}, t)\mathfrak{T}^{-1}\mathcal{P}^{-1}\mathcal{C}^{-1} = i[\psi^\dagger(-\mathbf{x}, -t)\gamma^5]_a = i[\gamma^5\psi^*(-\mathbf{x}, -t)]_a \tag{3.137}$$

For spin ½ tachyons

$$\mathcal{C}_T\mathcal{P}\mathcal{T}_T\psi_{Ta}(\mathbf{x}, t)\mathcal{T}_T^{-1}\mathcal{P}^{-1}\mathcal{C}_T^{-1} = -i[\psi^\dagger(-\mathbf{x}, -t)\gamma^5]_a \tag{3.138a}$$

where a is a spinor index. More succinctly,

$$\mathcal{C}_T\mathcal{P}\mathcal{T}_T\psi_T(\mathbf{x}, t)\mathcal{T}_T^{-1}\mathcal{P}^{-1}\mathcal{C}_T^{-1} = -i\gamma^5\psi^*(-\mathbf{x}, -t) \tag{3.138b}$$

Eq. 3.138 differs only by a phase from eq. 3.137.

Therefore one might think that bilinear combinations of ψ and $\psi^\dagger$ which of necessity must be factors in a Lagrangian will result in the cancellation of the –1 factors upon CPT transformation.

However the free field Lagrangian

$$\mathcal{L} = \psi_T^\dagger(x)i\gamma^0\gamma^5(\gamma^\mu\partial/\partial x^\mu + m)\psi_T(x) \tag{1.91}$$

changes to

$$\mathcal{L}_{CPT} = \psi_T^\dagger(-x)i\gamma^0\gamma^5(\gamma^\mu\partial/\partial x^\mu - m)\psi_T(-x) \tag{3.139}$$

The mass term violates CPT invariance. Thus massive spin ½ tachyons inherently violate CPT invariance. If all spin ½ particles start out massless and acquire masses through spontaneous symmetry breaking then the breaking of CPT invariance by massive, spin ½ tachyons is another consequence of spontaneous symmetry breaking.

Microcausality and Tachyons

Since the CPT Theorem does not hold for spin ½ tachyons it is of interest to consider to consider Jost's Theorem: CPT invariance is equivalent to weak local commutativity, which is a weak form of microcausality. In the case of spin ½ tachyons *weak local commutativity* (or weak microcausality) is defined as

$$<0|\{\psi_T^\dagger(x), \psi_T(y)\}|0> = 0 \quad \text{for} \quad (x-y)^2 < 0$$

(spacelike $(x-y)^2$).

The absence of CPT invariance leads us to inquire if microcausality, and/or weak microcausality, still hold?

To answer this question we evaluate the left-handed field commutator $\{\psi_{TL}^{+\dagger}(x), \psi_{TL}^{+}(y)\}$ to see if the normal microcausality condition holds:

$$\{\psi_{TL}^{+\dagger}(x), \psi_{TL}^{+}(y)\} = 0 \quad \text{for} \quad (x-y)^2 < 0 \tag{3.140}$$

We insert the Fourier expansions:

$$\{\psi_{TL}{}^{+}{}_{a}(x), \psi_{TL}{}^{+\dagger}{}_{b}(y)\} = \sum_{\pm s,s'} \int d^2p dp^{+} \int d^2p' dp'^{+}\, N_{TL}{}^{+}(p) N_{TL}{}^{+}(p')\theta(p^{+})\theta(p'^{+})\cdot$$
$$\cdot[\{b_{TL}{}^{+\dagger}(p',s'), b_{TL}{}^{+}(p,s)\} u_{TL}{}^{+}{}_{a}(p,s) u_{TL}{}^{+\dagger}{}_{b}(p',s') e^{+ip'\cdot y - ip\cdot x} +$$
$$+ \{d_{TL}{}^{+}(p',s'), d_{TL}{}^{+\dagger}(p,s)\} v_{TL}{}^{+}{}_{a}(p,s) v_{TL}{}^{+\dagger}{}_{b}(p',s') e^{-ip'\cdot y + ip\cdot x}]$$

$$= \sum_{\pm s} \int d^2p dp^{+}\, N_{TL}{}^{+2}(p)\theta(p^{+})[u_{TL}{}^{+}{}_{a}(p,s) u_{TL}{}^{+\dagger}{}_{b}(p,s) e^{+ip\cdot(y-x)} +$$
$$+ v_{TL}{}^{+}{}_{a}(p,s) v_{TL}{}^{+\dagger}{}_{b}(p,s) e^{-ip\cdot(y-x)}]$$

$$= i\int d^2p dp^{+}\theta(p^{+}) N_{TL}{}^{+2}(p)(2m|\mathbf{p}|)^{-1}\{[C^{-}R^{+}(-i\not{p} + m)\boldsymbol{\gamma}\cdot\mathbf{p}R^{+}C^{-}]_{ab} e^{+ip\cdot(y-x)} +$$
$$+ [C^{-}R^{+}(-i\not{p} - m)\boldsymbol{\gamma}\cdot\mathbf{p}R^{+}C^{-}]_{ab} e^{-ip\cdot(y-x)}\}$$

$$= \tfrac{1}{2}[C^{-}R^{+}]_{ab} \int d^2p_{\perp} \int_0^{\infty} dp^{+}(2\pi)^{-3}\,(e^{+ip\cdot(y-x)} + e^{-ip\cdot(y-x)})$$

where $p\cdot(y-x) = p^{-}(y^{+} - x^{+}) + p^{+}(y^{-} - x^{-}) - p_{\perp}\cdot(y_{\perp} - x_{\perp})$. Since $p^2 = -m^2$, the integral can be rewritten, after letting $p^{\mu} = -p^{\mu}$, as

$$= \tfrac{1}{2}[C^{-}R^{+}]_{ab} \int d^2p_{\perp} \int_{-\infty}^{\infty} dp^{+}(2\pi)^{-3}\, e^{-ip\cdot(y-x)}$$

where $p^{-} = (p_{\perp}{}^{2} + m^2)/(2p^{+})$. For spacelike $(x - y)^2 < 0$ we can always choose a coordinate system where $y^{-} - x^{-} = 0$ with the result

$$\{\psi_{TL}{}^{+\dagger}(x), \psi_{TL}{}^{+}(y)\} = 2^{-1}C^{-}R^{+}\,\delta(y^{-} - x^{-})\delta^2(\mathbf{y} - \mathbf{x}) \quad \underline{\text{if}} \quad y^{-} - x^{-} = 0 \quad (3.141)$$

Therefore

$$\{\psi_{TL}{}^{+\dagger}(x), \psi_{TL}{}^{+}(y)\} = 0 \quad \text{for} \quad (x - y)^2 < 0 \qquad (3.142)$$

Consequently, free left-handed (or right-handed) tachyons with light-front quantization satisfy the microcausality condition.

3.8 Perturbation Theory

If interactions are introduced, "normal particle" light-front perturbation theory has been shown to lead to the same results as conventional perturbation theory with the effects of the second term in eq. 3.29 cancelled in perturbation theory.[118]

Light-front perturbation theory can be extended to include interactions with tachyons by taking advantage of the locality and completeness of light-front tachyon fields, and by using the tachyon Feynman propagators developed earlier in this chapter.

---

[118] S-J Chang, R. Root, and T-M Yan, Phys. Rev. **D7**, 1133 (1973); S-J Chang, and T-M Yan, Phys. Rev. **D7**, 1147 (1973); T-M Yan, Phys. Rev. **D7**, 1761 (1973); T-M Yan, Phys. Rev. **D7**, 1780 (1973); J. Kogut and D. Soper, Phys. Rev. **D1**, 2901 (1970); J. D. Bjorken, J. Kogut, and D. Soper, Phys. Rev. **D3**, 1382 (1971); C. Thorn, Phys. Rev. **D19**, 639 (1979); and references therein.

Appendix B. Superluminal (Faster Than Light) Kinetic Theory and Thermodynamics

This Appendix contains two chapters on Superluminal Kinetic Theory and Thermodynamics that was published in 2012 in Blaha (2012a) and also in 2018 in Blaha (2018e).

This appendix deals with superluminal many particle dynamics. We will progress from superluminal Kinetic theory to Thermodynamics. We will see that there are strong similarities with non-relativistic Thermodynamics. This Appendix and Appendix A reinforce the view that superluminal Physics is perfectly acceptable theoretically. Its problem is our inability to dynamically transition from subluminal dynamics to superluminal dynamics due to the lack of "imaginary" complex forces.

B.1 Superluminal Kinetic Theory

Assemblages of large numbers of particles embody the Maxwell-Boltzmann distribution. The Boltzmann H theorem is the beginning point for derivations of the non-relativistic Maxwell-Boltzmann distribution. The non-relativistic Maxwell-Boltzmann distribution has the form

$$f(\mathbf{v}, \mathbf{r}) = n(m/(2\pi kT))^{3/2}\exp\{-[m(\mathbf{v} - \mathbf{v}_0)^2/2 + V(r)]/(kT)\} \tag{B.1}$$

where n is the particle density, T is the temperature, $\mathbf{v}_0$ is the average velocity, m is the particle mass, V(r) is an external conservative force, and k is Boltzmann's constant. In terms of a Hamiltonian

$$H(\mathbf{v}, \mathbf{r}) = m\mathbf{v}^2/2 + V(r) \tag{B.2}$$

we can express the Maxwell-Boltzmann distribution as

$$f(\mathbf{v}, \mathbf{r}) = n(m/(2\pi kT))^{3/2}\exp\{-H(\mathbf{v} - \mathbf{v}_0, \mathbf{r})/(kT)\} \tag{B.3}$$

B.1.1 Relativistic Form of the Maxwell-Boltzmann Distribution

If we assume that we have a container containing a distribution of relativistic (sublight) particles with an average velocity $\mathbf{v}_0 = 0$, and no external force, then the form of eq. B.3 generalizes to the relativistic Maxwell-Boltzmann distribution

$$f_R(\mathbf{v}) = C_R \exp\{-H/(kT)\} \tag{B.4}$$

where C_R is a normalization constant and H is the relativistic Hamiltonian for a free particle:

$$H = c(m^2c^2 + \mathbf{p}^2)^{1/2} \tag{B.5}$$

with $\mathbf{p} = \gamma m\mathbf{v}$ and $\gamma = (1 - v^2/c^2)^{-1/2}$. C_R is determined by the condition

$$\int d^3v f_R(\mathbf{v}) = 1 \tag{B.6}$$

B.1.2 Superluminal Form of the Maxwell-Boltzmann Distribution

The superluminal form of Maxwell-Boltzmann distribution is based on the form of the mass shell condition for superluminal particles:

$$E^2 - c^2\mathbf{p}^2 = m^2c^4 \tag{B.7}$$

which implies a free Hamiltonian

$$H_S = c(\mathbf{p}^2 - m^2c^2)^{1/2} \tag{B.8}$$

where

$$\mathbf{p} = \gamma_s m\mathbf{v} \tag{B.9}$$

and

$$\gamma_s = (v^2/c^2 - 1)^{-1/2}$$

The seemingly slight difference between eqs. B.8 and B.9, and eq. B.5 causes major differences between superluminal and relativistic kinetic theory and thermodynamics. On the other hand relativistic kinetic theory and thermodynamics are qualitatively similar in many ways with their non-relativistic counterparts.

One major difference is the behavior of kinematic variables near the speed of light:

<u>As v → c Below the Speed of Light</u>

$$p \to \infty$$
$$H \to \infty$$

<u>As v → c From Above the Speed of Light</u>

$$p \to \infty$$
$$H_S \to \infty$$

<u>As v → ∞</u>

$$p \to mc$$
$$H_S \to 0$$

Thus as v ranges from c to ∞, H_S decreases monotonically from ∞ to zero and p decreases from ∞ to mc. This behavior contrasts with H in eq. B.5, which increases monotonically with p as v increases from 0 to c. Thus the sublight Maxwell-Boltzmann distribution decreases with v as v increases from 0 to c.

The superluminal Maxwell-Boltzmann distribution *increases* with v as v increases from c to ∞ as we see below. The superluminal Maxwell-Boltzmann distribution decreases with p as p increases from mc to ∞. *As a result the natural physical parameterization of the Maxwell-Boltzmann distribution should be in terms of the momentum rather than the velocity.* Thus Boltzmann's H function which normally is

$$H_B(t) = \int d^3v\, f(\mathbf{v}, t) \log f(\mathbf{v}, t)$$

must be replaced with[119]

$$H_{BS}(t) = \int d^3p\, f_S(\mathbf{p}, t) \log f_S(\mathbf{p}, t)$$

The equilibrium superluminal Maxwell-Boltzmann distribution can be derived from $H_S(t)$. It has the same general form as the relativistic distribution

$$f_S(\mathbf{p}) = C_S \exp\{-H_S/(kT)\} \tag{B.10}$$

where C_S is a normalization constant and H_S is the superluminal Hamiltonian for a free particle.

We now apply the normalization condition[120]

$$n = N/V = \int d^3p f_S(\mathbf{p}) = C_S \int d^3p \exp\{-H_S/(kT)\} \tag{B.11}$$

where n is the particle density, N is the number of particles in the system, and V is the volume of the system. We calculate C_S by evaluating the integral:

$$n = 4\pi C_S \int_m^\infty dp\, p^2 \exp\{-H_S/(kT)\} \tag{B.12}$$

Letting $x = p/(mc)$ and $\alpha = mc^2/(kT)$ we see eq. B.12 becomes

$$n = 4\pi m^3c^3C_S \int_1^\infty dx\, x^2 \exp\{-\alpha(x^2 - 1)^{1/2}\} \tag{B.13}$$

Then letting $y^2 = x^2 - 1$ we find

$$n = 4\pi m^3c^3C_S \int_0^\infty dy\, y(y^2 + 1)^{1/2} \exp(-\alpha y)$$

$$= -m^3c^3C_S\, G^{31}_{13}(\alpha^2/4 \mid {}^{0}_{-3/2, 0, 1/2}) \tag{B.14}$$

---

[119] Note the additional factor of m^3 in $\int d^3p$ will be absorbed in the normalization (eq. H.11).

[120] We note that using $\int d^3v$ rather than $\int d^3p$ in eq. H.11 would result in a divergence – another reason for our choice of integration parameter.

where $G^{31}_{13}(\ldots)$ is Meijer's G-Function.[121] Therefore

$$C_S = -[m^3c^3\, G^{31}_{13}((mc^2/(2kT))^2 \,|\, {}^{0}_{-3/2,0,\,½})/n]^{-1} \tag{B.15}$$

The most probable momentum of a particle p_p is the maximum of

$$p_p = \text{Max}\{p^2\exp[-H_S/(kT)]\}$$

$$= \{(2(kT)^2/c^2)[1 + (1 - m^2c^4/(kT)^2)^{½}]\}^{½} \tag{B.16}$$

For large T or small T the maximum is

$$p_p \approx 2kT/c \;> mc$$

The velocity v_p corresponding to the maximum in the momentum is

$$v_p = cp_p/(p_p{}^2 - m^2c^2)^{½}$$

For large T or small T, the velocity v_p corresponding to the maximum in the momentum is approximately

$$v_p \approx c + ½\, m^2c^5/(2kT)^2$$

B.2 Superluminal Thermodynamics

Turning now to the thermodynamics of a dilute superluminal gas implied by the superluminal Maxwell-Boltzmann distribution we begin by calculating the average energy per particle

$$\varepsilon = C_S \int d^3p\, H_S \exp[-H_S/(2kT)] \Big/ \int d^3p\, C_S \exp[-H_S/(kT)] \tag{B.17}$$

$$= (C_S/n) \int d^3p\, H_S \exp[-H_S/(kT)]$$

$$= (C_S/n)2kT\alpha\; 4\pi m^2c^3 \int_0^\infty dy\; y^2(y^2 + 1)^{½} \exp(-\alpha y)$$

$$= -(C_S/n)m^3c^5\, G^{31}_{13}(\alpha^2/4 \,|^{-½}{}_{-2,0,\,½})$$

$$= mc^2\, G^{31}_{13}((mc^2/(2kT))^2|^{-½}{}_{-2,0,\,½})/G^{31}_{13}((mc^2/(2kT))^2|^{0}{}_{-3/2,0,\,½}) \tag{B.18}$$

The Maxwell-Boltzmann normalization factor is related to the energy per particle by

---

[121] See Gradshteyn (1965) integral 3.389.2 and p. 1068 for the properties of Meijer's G-Function.

$$C_S = -n\varepsilon/(m^3c^5\, G^{31}{}_{13}(\alpha^2/4\, |^{-1/2}{}_{-2,0,\, 1/2})) \qquad \text{(B.19)}$$

Note that C_S is proportional to the energy in contrast to the non-relativistic case where the Maxwell-Boltzmann normalization factor $C = (3m/(4\pi\varepsilon))^{3/2}$.

We now calculate the superluminal pressure for the case of a distribution of superluminal particles bouncing on a wall perpendicular to the z-axis. The wall is assumed to be a perfectly reflecting plane. The pressure is the average force per unit area due to the gas of superluminal particles. The number of particles bombarding the wall per second is with $v_z > 0$ is $v_z f_S(\mathbf{p})d^3p$. Thus the pressure is

$$P = \int d^3p\, 2p_z v_z f_S(\mathbf{p}) \qquad \text{(B.20)}$$

where the particle momentum changes by $2p_z$ due to reflection. Due to spherical symmetry one expects the average values for the various components of **v** to be equal. Consequently we can re-express eq. B.20 as

$$P = 1/3 \int d^3p\, 2m\gamma_s v^2 f_S(\mathbf{p}) \qquad \text{(B.21)}$$
$$= 1/3 \int d^3p\, 2pv f_S(\mathbf{p})$$

Since

$$v = cp/(p^2 - m^2c^2)^{1/2} \qquad \text{(B.22)}$$

we see

$$P = 8\pi c/3 \int_m^\infty dp\, p^4 f_S(\mathbf{p})/(p^2 - m^2c^2)^{1/2}$$

Following steps similar to eqs. B.12 – B.15 leads to

$$P = m^4c^4\, C_S\, G^{31}{}_{13}((mc^2/(2kT))^2|\, ^{1/2}{}_{-2,0,\, 1/2}) \qquad \text{(B.23)}$$

The *equation of state* relating the pressure and energy is

$$P = -(m/c)\{G^{31}{}_{13}((mc^2/(4kT))^2|\, ^{1/2}{}_{-2,0,\, 1/2})/G^{31}{}_{13}(\alpha^2/4\, |^{-1/2}{}_{-2,0,\, 1/2}))\}n\varepsilon \qquad \text{(B.24)}$$

Substituting for ε we find

$$P = -(nm^2c)\{G^{31}{}_{13}(\rho\, |\, ^{1/2}{}_{-2,0,\, 1/2})/\, G^{31}{}_{13}(\rho\, |^{0}{}_{-3/2,0,\, 1/2})\} \qquad \text{(B.25)}$$

where

$$\rho = (mc^2/(2kT))^2 \qquad \text{(B.26)}$$

Turning now to the consideration of a dilute gas the internal energy of the gas can be defined to be[122]

$$U(t) = N\varepsilon \tag{B.27}$$

We note that the work done by the superluminal gas if its volume increases by dV is PdV. Then the superluminal (and usual) form of the first law of thermodynamics is

$$dQ = dU + PdV \tag{B.28}$$

where Q is the heat absorbed. The heat capacity of the system for constant volume is

$$C_V = (\partial U/\partial T)_V \tag{B.29}$$

The second law of thermodynamics, Boltzmann's H theorem, is based on

$$H = -S/kV \tag{B.30}$$

where H is the negative of the entropy divided k times the volume V. In systems where there are no superluminal particles, the H theorem states that the entropy never decreases for an isolated gas of fixed volume.

We can calculate H for a superluminal system under equilibrium conditions, H_e, from[123]

$$H_e = \int d^3p f_S(\mathbf{p})\ln(f_S(\mathbf{p})) \tag{B.31}$$

$$= \int d^3p f_S(\mathbf{p})[\ln C_S - H_S/(kT)] \tag{B.32}$$
$$= n \ln C_S - \int d^3p f_S(\mathbf{p}) H_S/(kT)$$

$$= n \ln C_S - n\varepsilon/(kT) \tag{B.33}$$

by eqs. B.11 and B.17. Therefore

$$S = -kVH_{Se} = -kN \ln C_S + N\varepsilon/T \tag{B.34}$$

Consequently we obtain the superluminal *and* standard non-relativistic result

$$1/T = (\partial S/\partial U)_X \tag{B.35}$$

---

122 The internal energy of a gas of non-interacting non-relativistic particles is U(t) = 3NkT/2. In the superluminal case it appears that it is eq. H.18.

123 We consistently assume that integrals over the momentum $\int d^3p$ are the proper integration (rather than integrations over velocity $\int d^3v$) because, for example, the calculation of the normalization constant eq. H.11 would diverge if the integration were over $\int d^3v$.

where x represents all other extensive variables.

B.3 Approximate Calculation of Kinetic and Thermodynamic Quantities

We can obtain more tractable expressions for kinetic and thermodynamic quantities by assuming $\mathbf{p}^2 \gg m^2c^2$ and approximating the Hamiltonian (eq. B.8) with

$$H_{Sa} = cp \tag{B.36}$$

The approximate normalization condition is

$$n - N/V - \int d^3p f_{Sa}(\mathbf{p}) = C_{Sa}\int d^3p \exp\{\ H_{Sa}/(kT)\} \tag{B.37}$$

where n is the particle density, N is the number of particles in the system, and V is the volume of the system. C_S is determined by

$$n = 4\pi C_{Sa} \int_{mc}^{\infty} dp\, p^2 \exp\{-cp/(kT)\} \tag{B.38}$$

Letting $\alpha = c/(kT)$ we see eq. B.38 becomes

$$n = 4\pi C_{Sa}\, d^2/d\alpha^2 \int_{mc}^{\infty} dp \exp(-\alpha p)$$

$$= 4\pi C_{Sa}\, d^2/d\alpha^2\, [(1/\alpha)\exp(-\alpha mc)] \tag{B.39}$$

Therefore the normalization factor is

$$C_{Sa} = n/\{4\pi\, d^2/d\alpha^2\, [(1/\alpha)\exp(-\alpha mc)]\} \tag{B.40}$$

The most probable momentum of a particle p_p is the maximum of

$$p_{pa} = \mathrm{Max}\{p^2\exp[-H_{Sa}/(kT)]\}$$
$$= 2kT/c \tag{B.41}$$

The velocity v_{pa} corresponding to the maximum in the momentum is

$$v_{pa} = cp_{pa}/(p_{pa}^2 - m^2c^2)^{1/2}$$

For large T or small T, the velocity v_{pa} corresponding to the maximum in the momentum is approximately

$$v_{pa} \approx c + \tfrac{1}{2}\, m^2c^5/(2kT)^2$$

Turning now to the thermodynamics implied by the superluminal Maxwell-Boltzmann distribution we begin by calculating the average energy per particle

$$\varepsilon_a = \int d^3p\, H_{Sa} \exp[-H_{Sa}/(kT)]/\int d^3p \exp[-H_{Sa}/(kT)] \qquad \text{(B.42)}$$
$$= (C_{Sa}/n) \int d^3p\, H_{Sa} \exp[-H_{Sa}/(kT)]$$
$$= -(4\pi c C_{Sa}/n)\, d^3/d\alpha^3[(1/\alpha)\exp(-\alpha mc)] \rightarrow 3kT \text{ for } T \gg mc$$

where $\alpha = c/(kT)$.[124]

The Maxwell-Boltzmann normalization factor is related to the energy per particle by

$$C_{Sa} = -n\varepsilon_a/\{4\pi c\, d^3/d\alpha^3[(1/\alpha)\exp(-\alpha mc)]\} \qquad \text{(B.43)}$$

Note that C_{Sa} is proportional to the energy ε_a in contrast to the non-relativistic case where the Maxwell-Boltzmann normalization factor $C = (3m/(4\pi\varepsilon))^{3/2}$.

We now calculate the superluminal pressure for the case of a distribution of superluminal particles bouncing on a wall perpendicular to the z-axis. The wall is assumed to be a perfectly reflecting plane. The pressure is the average force per unit area due to the gas of superluminal particles. The number of particles bombarding the wall per second is with $v_z > 0$ is $v_z f_{Sa}(\mathbf{p})d^3p$. Thus the pressure is

$$P_a = \int d^3p\, 2p_z v_z f_{Sa}(\mathbf{p}) \qquad \text{(B.44)}$$

where the particle momentum changes by $2p_z$ due to reflection. Due to spherical symmetry one expects the average values for the various components of $\mathbf{v}$ to be equal. Consequently we can re-express eq. B.44 as

$$P_a = 1/3 \int d^3p\, 2m\gamma_s v^2 f_{Sa}(\mathbf{p}) \qquad \text{(B.45)}$$
$$= 1/3 \int d^3p\, 2pv f_{Sa}(\mathbf{p})$$

Since

$$v = cp/(p^2 - m^2c^2)^{1/2} \qquad \text{(B.46)}$$

we see

$$P_a = 8\pi c/3 \int_{mc}^{\infty} dp\, p^4 f_{Sa}(\mathbf{p})/(p^2 - m^2c^2)^{1/2} \qquad \text{(B.47)}$$

$$\cong 8\pi c/3 \int_{mc}^{\infty} dp\, p^3 f_{Sa}(\mathbf{p})$$

Evaluating eq. B.47 yields

$$P_a = -(8\pi c/3)\, C_{Sa}\, d^3/d\alpha^3[(1/\alpha)\exp(-\alpha mc)] \qquad \text{(B.48)}$$

The *equation of state* relating the pressure and energy is[125]

[124] The Superluminal case differs from the non-relativistic case: $\varepsilon_a = 3kT/2$. An example of $\varepsilon_a = 3kT$ is a crystal with a potential energy of compression. See p. 192 Morse (1964).
[125] The same equation of state as non-relativistic kinetic theory. See p. 72 Huang (1965).

$$P_a = 2/3\ n\varepsilon_a \quad (B.49)$$

For $T \gg mc$ we found[126]

$$\varepsilon_a \rightarrow 3kT \quad (B.50)$$

then, contrary to non-relativistic kinetic theory, we find ($T \gg mc$)

$$P_a = 2nkT \quad (B.51)$$

Turning now to the consideration of a dilute gas the internal energy of the gas for $T \gg mc$ is

$$U(t) = N\varepsilon \rightarrow 3NkT \quad (B.52)$$

We note again that the work done by the superluminal gas if its volume increases by dV is PdV. Then the superluminal (and usual) form of the first law of thermodynamics is

$$dQ = dU + PdV \quad (B.53)$$

where Q is the heat absorbed. The heat capacity of the system for constant volume is ($T \gg mc$)

$$C_V \rightarrow 3Nk \quad (B.54)$$

The second law of thermodynamics, Boltzmann's H theorem, is based on

$$H = -S/kV \quad (B.55)$$

where H is the negative of the entropy divided k times the volume V. In systems where there are no superluminal particles, the H theorem states that the entropy never decreases for an isolated gas of fixed volume.

We can calculate H_{BS} for a superluminal system under equilibrium conditions, H_{BSea}, from[127]

$$\begin{aligned} H_{BSea} &= \int d^3p f_{Sa}(\mathbf{p}) \ln(f_{Sa}(\mathbf{p})) && (B.56) \\ &= \int d^3p f_{Sa}(\mathbf{p})[\ln C_{Sa} - H_{Sa}/(kT)] && (B.57) \\ &= n \ln C_{Sa} - \int d^3p f_{Sa}(\mathbf{p}) H_{Sa}/(kT) \\ &= n \ln C_{Sa} - n\varepsilon_a/(kT) && (B.58) \end{aligned}$$

---

[126] Later we will define temperature in terms of the entropy S as $1/T = (\partial S/\partial U)_x$ where x is all other extensive variables.

[127] We consistently assume that integrals over the momentum $\int d^3p$ are the proper integration (rather than integrations over velocity $\int d^3v$) because, for example, the calculation of the normalization constant eq. H.11 would diverge if the integration were over $\int d^3v$.

by eqs. B.11 and B.17. Therefore

$$S_a = -kVH_{BSea} = -kN \ln C_{Sa} + N\varepsilon_a/T \qquad (B.59)$$

The superluminal *and* standard non-relativistic result still holds

$$1/T = (\partial S/\partial U)_x \qquad (B.60)$$

where x represents all other extensive variables.

B.4 Superluminal Kinetics and Thermodynamics Are Similar to the Non-Relativistic Case

In the previous sections we have shown that kinetic theory and the laws of thermodynamics are usually similar in the superluminal and non-relativistic cases modulo detail differences in the values of the various quantities due to differences between superluminal kinematics and non-relativistic kinematics.

Appendix C. Complex General Relativity and the U(4) Species Group

This Appendix considers Complex General Relativity and finds that it reduces to a U(4) representation combined with General Relativity for real-valued coordinates. The U(4) group may have a corresponding Internal Symmetry group that we have called the *Species Group*. Since the U(4) group has interactions with all fermions and bosons we find it leads to the known, but previously underived, principle

Inertial mass equals gravitational mass

This Appendix is from Blaha (2020c).

50. Complex General Relativity Reformulated

We have seen that Complex Special Relativity is the basis of flat space-time phenomena. Flat space-time coordinates are complex-valued in general. The real-valued coordinates that we experience in everyday life are the result of our measuring instruments: clocks and rulers. Real-valued coordinates are generated from complex-valued coordinates by Reality group transformations.

If flat space-time is governed by Complex Special Relativity then it is clear that curved 'space-time' is governed by Complex General Relativity. Here again there is a Reality group the General Relativistic Reality group – a U(4) group – that maps complex-valued General Relativity coordinates to real-valued curved coordinates.[128] There is a corresponding U(4) Internal Symmetry Reality group that we call the *Species group*. This group rotates fermions.

We can isolate the General Relativistic Reality group by factoring complex General Relativistic coordinate transformations into parts that consist of a real-valued General Coordinate transformation and complex-valued coordinate transformations. It will be apparent that the General Relativistic Reality group emerges in this discussion. The Species group is distinct from the Internal Symmetry Reality group of the SuperStandard Theory: We begin by defining the tetrad notation.

50.1 Tetrad (Vierbein) Formalism

The *vierbein* formalism begins with the Equivalence Principle that allows us to define an inertial coordinate system in the neighborhood of any point Z in space-time. We will use the notation $\varsigma^{\alpha}(Z)$ to denote the inertial coordinates at Z. We define a tetrad or vierbein as

[128] Much of this chapter appears in Blaha (2016h) and (2017a).

$$v^{\alpha}{}_{\mu}(x) = (\partial\varsigma^{\alpha}(x)/\partial x^{\mu})_{x=Z} \tag{50.1}$$

and, in a neighborhood of Z, we can invert the relation between ς and x to define an inverse

$$w^{\mu}{}_{\alpha}(x) = (\partial x^{\mu}(\varsigma)/\partial\varsigma^{\alpha})_{x=X} \tag{50.2}$$

such that

$$\begin{aligned} w^{\mu}{}_{\alpha}(x)v^{\alpha}{}_{\nu}(x) &= \delta^{\mu}{}_{\nu} \\ w^{\mu}{}_{\beta}(x)v^{\alpha}{}_{\mu}(x) &= \delta^{\alpha}{}_{\beta} \end{aligned} \tag{50.3}$$

In real General Relativity all *tetrads* are real-valued. In Complex General Relativity a *tetrad* $v^{\alpha}{}_{\mu}(x)$ is complex-valued.

The metric at a curved space-time point X is defined in terms of *tetrads* as

$$\begin{aligned} g_{\rho\sigma}(x) &= \eta_{\alpha\beta}\, v^{\alpha}{}_{\rho}(x)v^{\beta}{}_{\sigma}(x) \\ g^{\rho\sigma}(x) &= \eta^{\alpha\beta}\, w^{\rho}{}_{\alpha}(x)w^{\sigma}{}_{\beta}(x) \end{aligned} \tag{50.4}$$

The inverse of a *tetrad* transformation can also be expressed as

$$w_{\beta}{}^{\nu}(x) = v_{\beta}{}^{\nu}(x) = \eta_{\beta\alpha}g^{\nu\mu}(x)v^{\alpha}{}_{\mu}(x)$$

Then a *tetrad* and its inverse satisfy

$$v^{\alpha}{}_{\mu}(x)v_{\beta}{}^{\mu}(x) = \delta^{\alpha}{}_{\beta} \tag{50.5}$$

and

$$v^{\alpha}{}_{\mu}(x)v_{\alpha}{}^{\nu}(x) = \delta^{\nu}{}_{\mu}$$

There are two general types of space-time transformations that can be performed on a tetrad.

1. A complex-valued (possibly real-valued) General Relativistic coordinate transformation:

$$v'^{\alpha}{}_{\mu}(x) = \partial x^{\nu}/\partial x'^{\mu}\, v^{\alpha}{}_{\nu}(x)$$

2. A complex-valued, local *Lorentzian transformation*

$$v'^{\beta}{}_{\mu}(x) = \Lambda(x)^{\beta}{}_{\alpha}\, v^{\alpha}{}_{\mu}(x)$$

where $\Lambda(x)^{\beta}{}_{\alpha}$ is an element of a subset of the local Complex Lorentz Group.

The local Lorentzian transformations $\Lambda(x)^{\beta}{}_{\alpha}$ consist of local Lorentz transformations that are real-valued, and complex-valued Lorentz transformations. Both types of transformations satisfy the orthogonality condition:

$$\eta_{\alpha\beta}\Lambda^{\alpha}{}_{\rho}(x)\Lambda^{\beta}{}_{\sigma}(x) = \eta_{\rho\sigma} \qquad (50.6)$$

Thus the *tetrad* partakes of both local (position dependent) General Relativistic transformations and local Lorentzian transformations.

50.2 Complex General Relativistic Transformations

The General Relativistic Reality group interaction emerges from complex General Relativistic transformations. We can separate elements of the set of all complex General Coordinate transformations into a product of two factors: a real-valued General Coordinate transformation and a complex-valued General Coordinate transformation. The set of complex factors can be further factored into those that satisfy

$$\Lambda(\omega, \mathbf{u})^{T}G\Lambda(\omega, \mathbf{u}) = G \qquad (50.7)$$

and those that do not. We then see that the set of those that do not satisfy the above equation form a curved space representation of the U(4) group under 'multiplication' of transformations.

The elements of the set of real and complex General Coordinate transformations whose flat complex space-time limit satisfy the above equation form the elements of the Complex Lorentz group.[129]

We thus find the set of all 4-dimensional complex, curved space General coordinate transformations can be visualized as in Fig. 50.1. The next section describes the interplay of the three parts displayed in Fig. 50.1.

50.3 Structure of Complex General Coordinate Transformations

Complex General Coordinate transformations can be uniquely factored into products of two terms, which will later be further factored into three factors. They have the form

$$\partial x''^{\nu}(x)/\partial x^{\mu} = U(x'')^{\nu}{}_{\beta}\,\partial x'^{\beta}(x)/\partial x^{\mu} \qquad (50.8)$$

where

$$x''^{\nu}(x) = U(x'')^{\nu}{}_{\beta}x'^{\beta}$$
$$x'^{\mu}(x) = U^{-1\mu}{}_{b}(x'')\, x''^{b}$$

where $U(x')^{\nu}{}_{\beta}$ is complex and where $\partial x'^{\beta}(x)/\partial x^{\mu}$ is a purely real General Coordinate transformation.

We define

$$U\,(x'')^{\mu}{}_{\nu} = w^{\mu}{}_{a}(x'')[\exp(i\textstyle\sum_{k} g_{k}\Phi_{k}(x'')\tau_{k})]^{a}{}_{b}\, v^{b}{}_{\nu}(x'') \qquad (50.9)$$

---

[129] It is this part of curved space-time General Relativity that becomes the flat space-time Complex Lorentz group, which leads to the SU(3)⊗SU(2)⊗U(1)⊗SU(2)⊗U(1)⊗SU(3) Standard Model Reality group.

$$U^{-1}(x'')^{\mu}{}_{\nu} = w^{\mu}{}_{a}(x'')[\exp(-i\textstyle\sum_k g_k\Phi_k(x'')\tau_k)]^{a}{}_{b}\, v^{b}{}_{\nu}(x'')$$

where the constants g_k are real, and Φ_k and τ_k are hermitean. The uniqueness of the factorization follows from the Reality group (and U(4)) property that any complex 4-vector can be uniquely mapped to any specified real 4-vector.)

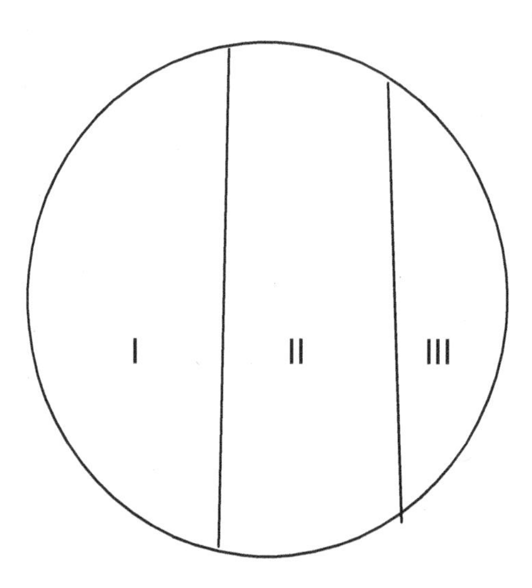

Figure 50.1. A visualization of the set of General Coordinate transformations separated into real-valued General coordinate transformations (part I), complex transformations that satisfy $\Lambda(\omega, u)^T G\Lambda(\omega, u) = G$ (part II), and complex transformations that do not satisfy $\Lambda(\omega, u)^T G\Lambda(\omega, u) = G$ (part III). Part I and part II combine in the limit of flat space-time to form the Complex Lorentz group. Parts II and III elements form a U(4) group that we call the General Relativistic Reality group.

Given the factorization above it becomes possible to separate the affine connection correspondingly.

50.4 Complex Affine Connection – General Relativistic Reality Group

The structure of a complex general coordinate transformation enables us to calculate its affine connection for later use in determining the covariant derivative, and the dynamic equations. First the transformation to the real-valued x' coordinates from inertial coordinates is

$$\Gamma^{\sigma}{}_{\lambda\mu}(x') = \partial x'^{\sigma}/\partial\varsigma^{\rho}\ \partial^2\varsigma^{\rho}/\partial x'^{\lambda}\partial x'^{\mu} \tag{50.10}$$

Next the Reality group transformation has the affine connection

$$\Gamma^{\sigma}{}_{\lambda\mu}(x'') = \partial x''^{\sigma}/\partial\varsigma^{\rho}\ \partial^2\varsigma^{\rho}/\partial x''^{\lambda}\partial x''^{\mu}$$

which can be re-expressed as

$$\Gamma^{\sigma}{}_{\lambda\mu}(x'') = \partial x''^{\sigma}/\partial x'^{\beta}\, \partial x'^{\beta}(\varsigma)/\partial\varsigma^{\rho}\, \partial/\partial x''^{\mu}[\partial\varsigma^{\rho}/\partial x'^{\alpha}\, \partial x'^{\alpha}/\partial x''^{\lambda}] =$$
$$= \partial x''^{\sigma}/\partial x'^{\beta}\, \partial x'^{\alpha}/\partial x''^{\lambda}\, \partial x'^{\gamma}/\partial x''^{\mu}\, \Gamma^{\beta}{}_{\alpha\gamma}(x') + \partial x''^{\sigma}/\partial x'^{\beta}\, \partial^2 x'^{\beta}/\partial x''^{\lambda}\partial x''^{\mu} \tag{50.11}$$

Next substituting the General Relativistic Reality group transformation

$$x''^{\nu}(x) = U(x'')^{\nu}{}_{\beta}x'^{\beta}$$
$$x'^{\mu}(x) = U^{-1}(x'')^{\mu}{}_{\beta}\, x''^{\beta}$$

together with

$$\partial x''^{\sigma}/\partial x'^{\beta} = \partial[U(x'')^{\sigma}{}_{\alpha}x'^{\alpha}]/\partial x'^{\beta} = U(x'')^{\sigma}{}_{\beta} + x'^{\alpha}\, \partial U(x'')^{\sigma}{}_{\alpha}/\partial x'^{\beta}$$

$$\partial x'^{\sigma}/\partial x''^{\beta} = \partial[U^{-1}(x'')^{\sigma}{}_{\alpha}x''^{\alpha}]/\partial x''^{\beta} = U^{-1}(x'')^{\sigma}{}_{\beta} + x''^{\alpha}\, \partial U^{-1}(x'')^{\sigma}{}_{\alpha}/\partial x''^{\beta}$$

we find the second term above is the Reality fields affine connection

$$\Gamma_R{}^{\sigma}{}_{\lambda\mu}(x'') = \partial[U(x'')^{\sigma}{}_{\alpha}x'^{\alpha}]/\partial x'^{\beta}\, \partial\{\partial[U^{-1}(x'')^{\beta}{}_{\alpha}x''^{\alpha}]/\partial x''^{\lambda}\}/\partial x''^{\mu}$$

and so we find the affine connections are approximately additive. Thus approximately

$$\Gamma^{\sigma}{}_{\lambda\mu}(x'') = \Gamma_{GR}{}^{\sigma}{}_{\lambda\mu}(x') + \Gamma_R{}^{\sigma}{}_{\lambda\mu}(x'')$$

if $x''^{\sigma} \simeq x'^{\sigma}$.

A complex transformation of types II and III in Fig. 50.1 has the form:

$$U(x'')^{\mu}{}_{\nu} = w^{\mu}{}_{a}(x'')[\exp(i\textstyle\sum_k \Phi_k(x'')\tau_k)]^{a}{}_{b}\, 1^{b}{}_{\nu}(x'')$$
$$U^{-1}(x'')^{\mu}{}_{\nu} = w^{\mu}{}_{a}(x'')[\exp(-i\textstyle\sum_k \Phi_k(x'')\tau_k)]^{a}{}_{b}\, 1^{b}{}_{\nu}(x'')$$

where τ_k is a U(4) generator matrix. Its infinitesimal transformation is approximately

$$U(x'')^{\nu}{}_{\beta} \approx \delta^{\nu}{}_{\beta} + i\textstyle\sum_k \Phi_k(x'')[\tau_k]^{\nu}{}_{\beta} \tag{50.12}$$
$$U^{-1}(x'')^{\nu}{}_{\beta} \approx \delta^{\nu}{}_{\beta} - i\textstyle\sum_k \Phi_k(x'')[\tau_k]^{\nu}{}_{\beta}$$

using the *vierbein* flat space-time limits

$$w^{\mu}{}_{a}(x'') \approx \delta^{\mu}{}_{a}$$
$$1^{b}{}_{\nu}(x'') \approx \delta^{b}{}_{\nu}$$

where

$$\Phi_k(x) = \int^{x} dy_{\lambda}\, A_{Rk}{}^{\lambda}(y) \tag{50.13}$$

Then

$$\Gamma_R{}^{\sigma}{}_{\lambda\mu} = -\tfrac{1}{2}i\{\textstyle\sum_k A_{Rk}(x'')_{\mu}[\tau_k]^{\sigma}{}_{\lambda} + \sum_k A_{Rk}(x'')_{\lambda}[\tau_k]^{\sigma}{}_{\mu}\} \tag{50.14}$$

$$= A_R{}^{\sigma}{}_{\mu\lambda} + A_R{}^{\sigma}{}_{\lambda\mu}$$

(summed over k) with the matrix $A_R{}^{\sigma}{}_{\mu\lambda}$ given by

$$A_R{}^{\sigma}{}_{\mu\lambda} = -\tfrac{1}{2}i\sum_k A_{Rk\mu}[\tau_k]^{\sigma}{}_{\lambda} \tag{50.15}$$

with $A_R{}^{\sigma}{}_{\mu\lambda}$ transformable to matrix row and column numbers

$$A_{R_{flat}}{}^{\mu a}{}_{b} = A_{R_{flatk}}{}^{\mu}[\tau_k]^{\sigma}{}_{\lambda}\delta_{\sigma}{}^{a}\delta^{\lambda}{}_{b}$$

using the flat space-time vierbein values, and so $A_{R_{flat}}{}^{a}{}_{\mu b}$ may be written in matrix form as

$$A_{R_{flat\mu}} = -\tfrac{1}{2}i\sum_k A_{R_{flatk\mu}}\tau_k \tag{50.16}$$

In the flat space-time limit the $A_{Rk}{}^{\lambda}(y)$ becomes the Coordinate Species group U(4) gauge fields $A_{R_{flatk}}{}^{\lambda}(y)$.

The relevant *quadratic* $A_R{}^{\sigma}{}_{\mu\lambda}$ terms from eq. 22.22 below that are needed to find the dynamic equation for the gauge fields $A_{R_{flat}}{}^{i}{}_{\mu}$ are contained in

$$\mathcal{L}_A = \mathrm{Tr}\,\sqrt{g}[M\partial_\nu R^1{}_{\sigma\mu}\partial^{\nu}R^{2\sigma\mu} + aR^1{}_{\sigma\mu}R^{2\sigma\mu} + bg^{\sigma\mu}(R^1{}_{\sigma\mu} + R^2{}_{\sigma\mu}) + 1/4(g_{\mu\nu} + g^2{}_{\mu\nu})T^{\mu\nu}] \tag{50.17}$$

We can let

$$R^i{}_{\sigma\mu} = R^{i\beta}{}_{\sigma\beta\mu} \equiv \partial_\mu(A_R{}^{i\beta}{}_{\sigma\beta} + A_R{}^{i\beta}{}_{\beta\sigma}) - \partial_\beta(A_R{}^{i\beta}{}_{\sigma\mu} + A_R{}^{i\beta}{}_{\mu\sigma}) \tag{50.18}$$

for i = 1, 2. In the flat space-time limit we chose the Landau gauge

$$\partial_\mu A_{R_{flat}}{}^{i\mu a}{}_{b} = 0 \tag{50.19}$$

As a result

$$R^i{}_{\sigma\mu} \equiv \partial_\mu(A_R{}^{i\beta}{}_{\sigma\beta} + A_R{}^{i\beta}{}_{\beta\sigma}) \tag{50.20}$$

Using

$$A_{R_{flat}}{}^{i\mu\sigma}{}_{\lambda} = A_{R_{flat}}{}^{i\mu}[\tau_k]^{a}{}_{b}\delta^{\sigma}{}_{a}\delta_{\lambda}{}^{b} \tag{50.21}$$
$$A_{R_{flat}}{}^{i}{}_{\mu} = -\tfrac{1}{2}i\sum_k A_{R_{flat}}{}^{i}{}_{k\mu}$$

and taking the trace in eq. 50.17 we obtain

$$\mathcal{L}_A = \mathrm{Tr}\,\sqrt{g}[8M\partial_\nu\partial_\mu A_{R_{flat}}{}^{1}{}_{\sigma}\partial^{\nu}\partial^{\mu}A_{R_{flat}}{}^{2\sigma} + 8a\partial_\mu A_{R_{flat}}{}^{1}{}_{\sigma}\partial^{\mu}A_{R_{flat}}{}^{2\sigma} + 1/4(g_{\mu\nu} + g^2{}_{\mu\nu})T^{\mu\nu}] \tag{50.22}$$

in the flat space-time limit. Eq. 50.17 needs to take account of the complex nature of $(g_{\mu\nu} + g^2_{\mu\nu})$ until transformed by the infinitesimal form of the complex Reality transformation:

$$\begin{aligned}(g_{\beta\alpha} + g^2_{\beta\alpha})' &\rightarrow U(x'')_\beta{}^\mu(g_{\mu\nu} + g^2_{\mu\nu})U^{-1}(x'')^\nu{}_\alpha \\ &= (\delta_\beta{}^\mu + i\sum_k \Phi_k(x'')[\tau_k]_\beta{}^\mu)(g_{\mu\nu} + g^2_{\mu\nu})(\delta^\nu{}_\alpha - i\sum_k \Phi_k(x'')[\tau_k]^\nu{}_\alpha) \\ &\cong (g_{\beta\alpha} + g^2_{\beta\alpha}) + i\{\sum_k \Phi_k(x'')[\tau_k]_\beta{}^\mu - i\sum_k \Phi_k(x'')[\tau_k]^\nu{}_\alpha)(g_{\mu\nu} + g^2_{\mu\nu}) \\ &\cong (g_{\beta\alpha} + g^2_{\beta\alpha}) + i\sum_k \Phi_k(x'')\{[\tau_k]_{\beta\alpha} - [\tau_k]_{\beta\alpha}\} \end{aligned} \quad (50.23)$$

Approximating $\Phi_k(x'')$ with an infinitesimal line we find

$$\sum_k \Phi_k(x) \cong \delta x_\lambda A_{R\text{flat}}{}^{1\lambda}(x) \quad (50.24)$$

by eq. 50.13. Thus

$$1/4(g_{\mu\nu} + g^2_{\mu\nu})T^{\mu\nu} \cong \tfrac{1}{4}\,[(g_{\beta\alpha} + g^2_{\beta\alpha}) + i\,\delta x_\lambda A_{R\text{flat}}{}^{1\lambda}(x)\{[\tau_k]_{\beta\alpha} - [\tau_k]_{\beta\alpha}\}]T^{\alpha\beta}$$

Applying the canonical Euler-Lagrange method we obtain the dynamical equations (using integration by parts to handle higher order derivative terms):[130]

$$\Box^2 A_{R_{\text{flat}}}{}^1{}_\sigma + (a/M)\Box A_{R_{\text{flat}}}{}^1{}_\sigma + i\delta x_\sigma\{[\tau_k]_{\beta\alpha} - [\tau_k]_{\beta\alpha}\}T^{\alpha\beta}/(32M) = 0 \quad (50.25)$$

$$\Box^2 A_{R_{\text{flat}}}{}^2{}_\sigma + (a/M)\Box A_{R_{\text{flat}}}{}^2{}_\sigma = 0 \quad (50.26)$$

Since the $A_{R_{\text{flat}}}$ gauge field is gravitational in nature it exists, as eq. 50.25 shows, as a type of gravitational interaction whose source is the energy-momentum tensor. Following the standard derivation of the gravitational potential we find the Coulomb interaction of $A_{R_{\text{flat}}}$[10].

50.5 Species Interaction Gravity Potential

Assuming that we are dealing with non-relativistic matter we can calculate the gravity potential contribution from section 54.4 below:[131]

$$V_{GA1}(\mathbf{x}) = -(1/32M)\int d^3k\, \exp(i\mathbf{k}\cdot\mathbf{x})V_{GA1}(\mathbf{k})/(2\pi)^3 \quad (50.27)$$

where

---

[130] It is possible that the Reality transformation also depends on $A_{R_{\text{flat}}}{}^2{}_\sigma$. Then eq. 50.26 would have an energy-momentum tensor term as well. Consequently there would be an additional interaction of the same form as in eq. 50.27 below.

[131] **In absence of Higgs breaking. Later we add Higgs Species breaking terms that further increases the masses of the Species gauge fields. The addition of Higgs mass contributions to Species gauge fields does not conflict with the mass term found here but merely adds further contributions to Species gauge field masses. These contributions change these masses but would not appear to significantly change their order of magnitude.**

$$V_{GA1}(\mathbf{k}) = (\mathbf{k}^4 + (a/M)\mathbf{k}^2)^{-1} \quad (50.28)$$

The eq. 50.28 can be separated into two terms:

$$V_{GA1}(\mathbf{k}) = (M/a)[1/k^2 - 1/(\mathbf{k}^2 + a/M)] \quad (50.29)$$

which yield

$$V_{GA1}(\mathbf{r}) = -[1/(96\pi a)][1/r - e^{-m_A r}/r] \quad (50.30)$$

Later in section 55.3 and 55.5 we will see

$$M \sim 1.46\times10^{179}\ \text{GeV}^{-2} \quad (50.31)$$

$$m_A = (a/M)^{1/2} = 2.49\times10^{-90}\ \text{GeV} \quad (50.32)$$

Since $a \cong 1$ the coupling constant

$$1/(96\pi a) \cong 0.0033 \quad (50.33)$$

In comparison the electromagnetic fine structure constant is

$$\alpha \cong 0.0073$$

Thus the Species A_R coupling constant is approximately ½ of the fine structure constant.

50.6 Influence of Gravitational Gauge Field on Gravitation

The gravitational gauge field potential in eq. 50.30 has a relatively large coupling constant that makes it competitive with the known force of gravity at large distances of the scale of galactic distances. This force, which is negligible at short distances of the order of planetary distances, may be part of the MoND phenomena that affects the motion of stars.

50.7 PseudoQuantization of Affine Connections

Having obtained the form of the general affine connection we now PseudoQuantize them for later use in our unification program. We define

$$\begin{aligned} R^{1\beta}{}_{\sigma\nu\mu} &= \partial_\mu H^{1\beta}{}_{\sigma\nu} - \partial_\nu H^{1\beta}{}_{\sigma\mu} + H^{1\gamma}{}_{\nu\sigma} H^{1\beta}{}_{\gamma\mu} - H^{1\gamma}{}_{\mu\sigma} H^{1\beta}{}_{\gamma\nu} \\ R^{2\beta}{}_{\sigma\nu\mu\rho} &= \partial_\mu H^{2\beta}{}_{\sigma\nu} - \partial_\nu H^{2\beta}{}_{\sigma\mu} + H^{2\gamma}{}_{\nu\sigma} H^{2\beta}{}_{\gamma\mu} - H^{2\gamma}{}_{\mu\sigma} H^{2\beta}{}_{\gamma\nu} + \\ &\quad + H^{1\gamma}{}_{\nu\sigma} H^{2\beta}{}_{\gamma\mu} - H^{1\gamma}{}_{\mu\sigma} H^{2\beta}{}_{\gamma\nu} + H^{2\gamma}{}_{\nu\sigma} H^{1\beta}{}_{\gamma\mu} - H^{2\gamma}{}_{\mu\sigma} H^{1\beta}{}_{\gamma\nu} \end{aligned} \quad (50.34)$$

where

$$H^{\sigma}{}_{\nu\mu} = \Gamma_{GR}{}^{\sigma}{}_{\nu\mu} + \Gamma_{GR}{}^{2\sigma}{}_{\nu\mu} + \Gamma_{R}{}^{1\sigma}{}_{\nu\mu} + \Gamma_{R}{}^{2\sigma}{}_{\nu\mu} \quad (50.35)$$

and where $\Gamma_{GR}{}^{\sigma}{}_{\nu\mu}$ and $\Gamma_{GR}{}^{2\sigma}{}_{\nu\mu}$ are affine connections for real-valued General Relativity, and $\Gamma_{R}{}^{1\sigma}{}_{\nu\mu}$ and $\Gamma_{R}{}^{2\sigma}{}_{\nu\mu}$ are affine connections for a complex-valued set of transformations embodying a U(4) gauge group that combine with real-valued General Relativistic transformations to yield Complex General Relativistic transformations.

The affine connection is most often viewed as a derived quantity—part of the derivation of the curvature tensor in General Relativity. It is typically derived from manipulations of the metric $g_{\mu\nu}$. However, the affine connection can also be viewed as a set of independent fields that become related to the metric via dynamic equations.

Some years ago A. Einstein and H. Weyl[132] pointed out that the metric and the affine connection should be treated as independent quantities and subject to independent arbitrary infinitesimal variations:

> "In contrast to Einstein's original "metric" conception in terms of the $g_{\nu\mu}$ there was later developed, by Eddington, by Einstein himself, and recently by Schrödinger, an affine field theory operating with the components $\Gamma^{\sigma}{}_{\nu\mu}$ of an affine connection. But in 1925 Einstein also advocated a "mixed" formulation by means of a Lagrangian in which both the $g_{\nu\mu}$ and the $\Gamma^{\sigma}{}_{\nu\mu}$ are taken as basic field quantities and submitted to independent arbitrary infinitesimal variations.[133] In certain respects this seems to be the most natural procedure."

Following this approach we have introduced the above affine connections for use in the construction of our unification of particle interactions.

[132] H. Weyl, Phys. Rev. **77**, 699 (1950).
[133] A. Einstein, Sitzungsber., Preuss. Akad. Der Wissensch. (1925), p. 414.

51. Species Group U(4) Gauge Fields

From the discussion in sections 50.4 - 50.5 we see the flat space-time limit of $A_{Rk}{}^{\lambda}(y)$ is a local U(4) coordinate space gauge field. There are, *by assumption*,[134] a corresponding internal symmetry gauge fields $A_{Sk}{}^{\lambda}(y)$ – the Internal Symmetry U(4) Species Group gauge fields. The mathematical features of this field is quite similar to the U(4) Generation group fields. The interaction that appears in covariant derivatives is $g_8\mathbf{A}_S{}^{\mu}(x) = g_8\mathbf{A}_{Sk}{}^{\mu}(x)\mathbf{G}_{Sk}$ where the $\mathbf{G}_{Sk}$ are U(4) generator matrices and k is summed from 1, … , 16.

Below we will see that the effect of the Internal Symmetry Species Group is to U(4) rotate the four components of each fermion's field. Since it preserves the species of each fermion we call this group the *Species Group*. It performs a U(4) rotation of the spinor representation of each fermion

We will see that the Higgs Mechanism breakdown of the Species Group endows each fermion with a mass contribution that breaks the scale invariance of the Unified SuperStandard Theory.

Since the Species Group Higgs Mechanism breaking gives each fermion a 'gravity generated' mass, and since this mass sets the mass scale for each fermion, we conclude later that the principle of the equality of inertial and gravitational mass is a direct consequence. ***Inertial mass equals gravitational mass.***

51.1 Species Group Covariance

A Species Group transformation on a Dirac equation must be covariant. Consider the Dirac equation Lagrangian term under an Internal Symmetry Species Group transformation:

$$\bar{\psi}(x)[i\gamma_\mu(\partial/\partial x_\mu - ig_8A_{Sk}{}^{\mu}(x)G_{Sk}) - m]\psi(x) = 0 \qquad (51.1)$$

summed over k. If we perform a Species group transformation U on Lagrangian terms:

$$\bar{\psi}(x)[i\gamma_\mu(\partial/\partial x_\mu - ig_8A_{Sk}{}^{\mu}(x)G_{Sk}) - m]U^{-1}U\psi(x)$$

or

$$\bar{\psi}(x)U^{-1}U[iU^{-1}U\gamma_\mu U^{-1}U(\partial/\partial x_\mu - ig_8A_{Sk}{}^{\mu}(x)G_{Sk}) - m]U^{-1}U\psi(x)$$

we find

$$\bar{\psi}'(x)[i\gamma_\mu'U(\partial/\partial x_\mu - ig_8A_{Sk}{}^{\mu}(x)G_{Sk})U^{-1} - m]\psi'(x)$$

---

[134] In this discussion we *assume* that the Coordinate Species Reality Group with gauge fields $A_{Rk}{}^{\lambda}(y)$ has a corresponding Internal Symmetry Reality group that we call the Internal Symmetry Species Group. This assumption parallels the assumptions for the SU(3)⊗SU(2)⊗U(1)⊗SU(2)⊗U(1) Internal Symmetry Reality Group presented in previous chapters.

where

$$\gamma_\mu'(x) = U\gamma_\mu U^{-1}$$

is locally equivalent to a Dirac matrix by Good's Theorem.[135] If we set

$$A'_S{}^\mu(x) = -(i/g_8)U[\partial U^{-1}/\partial x^\mu] + UA_S{}^\mu(x)U^{-1}$$

then the transformed Lagrangian terms are

$$\bar{\psi}'(x)[i\gamma_\mu'(x)(\partial/\partial x_\mu - ig_8A'_{Sk}{}^\mu(x)G_{Sk}) - m]\psi'(x) \qquad (51.2)$$

They have the same form as the original terms above and thus the expression is covariant. We note the indices of the matrices G_{Sk} are spinor indices and so $G_{Sk}\gamma_\mu$ has an implicit spinor matrix summation. But the symmetry group is U(4).

The coordinate dependence of $\gamma_\mu'(x)$ introduces locality into the Dirac matrix. This locality might be viewed with concern except that an inverse Species group transformation exists that removes the locality. Thus the physical impact of this 'new' locality is eliminated.

51.2 Spontaneous Symmetry Breaking of the General Relativity U(4) Reality Group – The Species Group

We begin the discussion of the Internal Symmetry Species Group symmetry breaking[136] by defining a Higgs field η which is a Species group 4-vector

$$\eta = \begin{bmatrix} \rho_1 \\ \rho_2 \\ \rho_3 \\ \rho_4 \end{bmatrix} \qquad (51.3)$$

where ρ_1, ρ_2, ρ_3 and ρ_4 are real fields.[137] Then the covariant derivative of η (taking account only of the Species group) is

$$D_{\ldots\mu}\eta = \{\partial/\partial X^\mu + \ldots - \tfrac{1}{2}ig_8\Sigma A_{Sk}{}^\mu(x)G_{Sk}\} \begin{bmatrix} \rho_1 \\ \rho_2 \\ \rho_3 \\ \rho_4 \end{bmatrix} \qquad (51.3)$$

[135] R. H. Good, Jr., Rev. Mod. Phys., **27**, 187 (1955).

[136] Since the Species gauge fields have been shown to have a mass it might seem redundant to introduce Higgs symmetry breaking as well. However the Higgs breaking introduces the further benefit of giving a mass term to each particle – thus establishing the equality of gravitational mass and inertial mass as we discuss in section 20.4.

[137] Each field ρ_i can be expressed as a PseudoQuantum field: $\rho_i = \varphi_{1i} + \varphi_{2i}$ where φ_{1i} has the vacuum expectation value ρ_{i0} for $i = 1, \ldots, 4$. Thus our PseudoQuantum field theory version is implemented easily.

Following steps similar to eqs. 47.4 through 47.17 for the Generation Group symmetry breaking we find with ρ_i being the vacuum expectation value of the Higgs field:

$$(D_{\dots\mu}\eta)^\dagger D_{\dots}^{\ \mu}\eta = \partial\rho_1/\partial X^\mu \partial\rho_1/\partial X_\mu + \partial\rho_2/\partial X^\mu \partial\rho_2/\partial X_\mu + \partial\rho_3/\partial X^\mu \partial\rho_3/\partial X_\mu +$$
$$+ \partial\rho_4/\partial X^\mu \partial\rho_4/\partial X_\mu +$$
$$+ \tfrac{1}{4} g_8^2\{\rho_1^2 A_{S1}^2 + \rho_2^2 A_{S2}^2 + \rho_3^2 A_{S3}^2 + \rho_4^2 A_{S4}^2 +$$
$$+ (\rho_1^2 + \rho_2^2)(V_5^2 + V_6^2) + \tfrac{1}{4}(\rho_1^2 + \rho_3^2)(V_7^2 + V_8^2) +$$
$$+ (\rho_1^2 + \rho_4^2)(V_9^2 + V_{10}^2) + \tfrac{1}{4}(\rho_2^2 + \rho_3^2)(V_{11}^2 + V_{12}^2) +$$
$$+ (\rho_2^2 + \rho_4^2)(V_{13}^2 + V_{14}^2) + \tfrac{1}{4}(\rho_3^2 + \rho_4^2)(V_{15}^2 + V_{16}^2)\} \tag{51.4}$$

up to total divergences, which generate surface terms which we discard. We also assume that all fields satisfy the gauge condition

$$\partial A_{Si}^{\ \mu}/\partial X^\mu = 0 \tag{51.5}$$

Eq. 51.4 shows all Species Group gauge fields have masses. Thus Species Group symmetry is completely broken. The combination of an ultra-weak coupling constant and very large gauge field masses results in extremely Species interactions.

We assume Species group gauge field masses to be very large – of the order of the Planck mass in view of its origin in Complex General Relativity.

51.3 Species Group Higgs Mechanism Contributions to Fermion Masses

The symmetry breaking of the Species Group results in a contribution to each fermion mass of all types, species, generations, and layers. The Species Group contributions to normal and Dark fermion mass terms are

$$\mathcal{L}^{Higgs}_{FermionMassesSpecies} = \Sigma_{s,g,l}\bar{\psi}_{sglL}\rho_s m_{sgl}\psi_{sglR} + \Sigma_{s,g,l}\bar{\psi}_{DsglL}\rho_s m_{Dsgl}\psi_{DsglR} + \text{c.c.} \tag{51.6}$$

The η field expectation value has components labeled ρ_s.[138] The mass matrices m_{sgl} and m_{Dsgl} are the complex constant Species mass matrix contributions for normal and Dark species.

51.4 Species Group Higgs Masses Shows Inertial Mass Equals Gravitational Mass

In Blaha (2016h) we showed that a Complex General Relativity transformation can be factored into the product of a complex-valued transformation and a real-valued General Coordinate transformation. The set of complex valued transformations form a U(4) group that we called the General Coordinate Reality group. The analogous Internal Symmetry Species Group has gauge fields that undergo spontaneous symmetry breaking and generate contributions to all fermion masses.

---

[138] The Higgs fields η… in our PseudoQuantum formulation are $\eta\ldots = \varphi_{1\ldots}(x) + \varphi_{2\ldots}(x)$ as described earlier.

Since fermion field masses are now sums of ElectroWeak Higgs contributions, Generation group Higgs contributions, Layer group Higgs contributions, and General Coordinate Species Group contributions, and since the Species Group Higgs fields appear in all fermion masses, the equality of inertial and gravitational mass is proven. The Species Group Higgs particles' equations set the mass scale of gravitational mass, and thereby of all Higgs mass contributions. The scale of inertial masses equals to the scale of gravitational masses. ***Since an expression cannot mix mass scales, the gravitational mass scale must be the same as the inertial mass scale.***

Inertial mass equals gravitational mass.

We have established the equality of inertial and gravitational mass at the short distance quantum level. In our view, this explanation is far more satisfying than basing the equality on a combination of large distance phenomena and quantum phenomena. As Einstein and Weyl have pointed out, all fundamental physics phenomena should be based on a local theory.

We have mapped Complex General Relativistic transformations consisting of U(4) transformations and Real General Relativistic transformations into transformations consisting of Internal Symmetry Species Group factors and Real General Relativistic transformations factors. The Higgs Mechanism breakdown of the Species Group has the important consequence that it prevents Species Group transformations that rotate between fermions and anti-fermions.

REFERENCES

Akhiezer, N. I., Frink, A. H. (tr), 1962, *The Calculus of Variations* (Blaisdell Publishing, New York, 1962).

Bjorken, J. D., Drell, S. D., 1964, *Relativistic Quantum Mechanics* (McGraw-Hill, New York, 1965).

Bjorken, J. D., Drell, S. D., 1965, *Relativistic Quantum Fields* (McGraw-Hill, New York, 1965).

Blaha, S., 1995, *C++ for Professional Programming* (International Thomson Publishing, Boston, 1995).

_______, 1998, *Cosmos and Consciousness* (Pingree-Hill Publishing, Auburn, NH, 1998 and 2002).

_______, 2002, *A Finite Unified Quantum Field Theory of the Elementary Particle Standard Model and Quantum Gravity Based on New Quantum Dimensions™ & a New Paradigm in the Calculus of Variations* (Pingree-Hill Publishing, Auburn, NH, 2002).

_______, 2004, *Quantum Big Bang Cosmology: Complex Space-time General Relativity, Quantum Coordinates™ Dodecahedral Universe, Inflation, and New Spin 0, ½, 1 & 2 Tachyons & Imagyons* (Pingree-Hill Publishing, Auburn, NH, 2004).

______, *2005a, Quantum Theory of the Third Kind: A New Type of Divergence-free Quantum Field Theory Supporting a Unified Standard Model of Elementary Particles and Quantum Gravity based on a New Method in the Calculus of Variations* (Pingree-Hill Publishing, Auburn, NH, 2005).

______, 2005b, *The Metatheory of Physics Theories, and the Theory of Everything as a Quantum Computer Language* (Pingree-Hill Publishing, Auburn, NH, 2005).

______, 2005c, *The Equivalence of Elementary Particle Theories and Computer Languages: Quantum Computers, Turing Machines, Standard Model, Superstring Theory, and a Proof that Gödel's Theorem Implies Nature Must Be Quantum* (Pingree-Hill Publishing, Auburn, NH, 2005).

______, 2006a, *The Foundation of the Forces of Nature* (Pingree-Hill Publishing, Auburn, NH, 2006).

______, 2006b, *A Derivation of ElectroWeak Theory based on an Extension of Special Relativity; Black Hole Tachyons; & Tachyons of Any Spin.* (Pingree-Hill Publishing, Auburn, NH, 2006).

_______, 2007a, *Physics Beyond the Light Barrier: The Source of Parity Violation, Tachyons, and A Derivation of Standard Model Features* (Pingree-Hill Publishing, Auburn, NH, 2007).

______, 2007b, *The Origin of the Standard Model: The Genesis of Four Quark and Lepton Species, Parity Violation, the ElectroWeak Sector, Color SU(3), Three Visible Generations of Fermions, and One Generation of Dark Matter with Dark Energy* (Pingree-Hill Publishing, Auburn, NH, 2007).

______, *2008a, A Direct Derivation of the Form of the Standard Model From GL(16) (Pingree-Hill Publishing, Auburn, NH, 2008).*

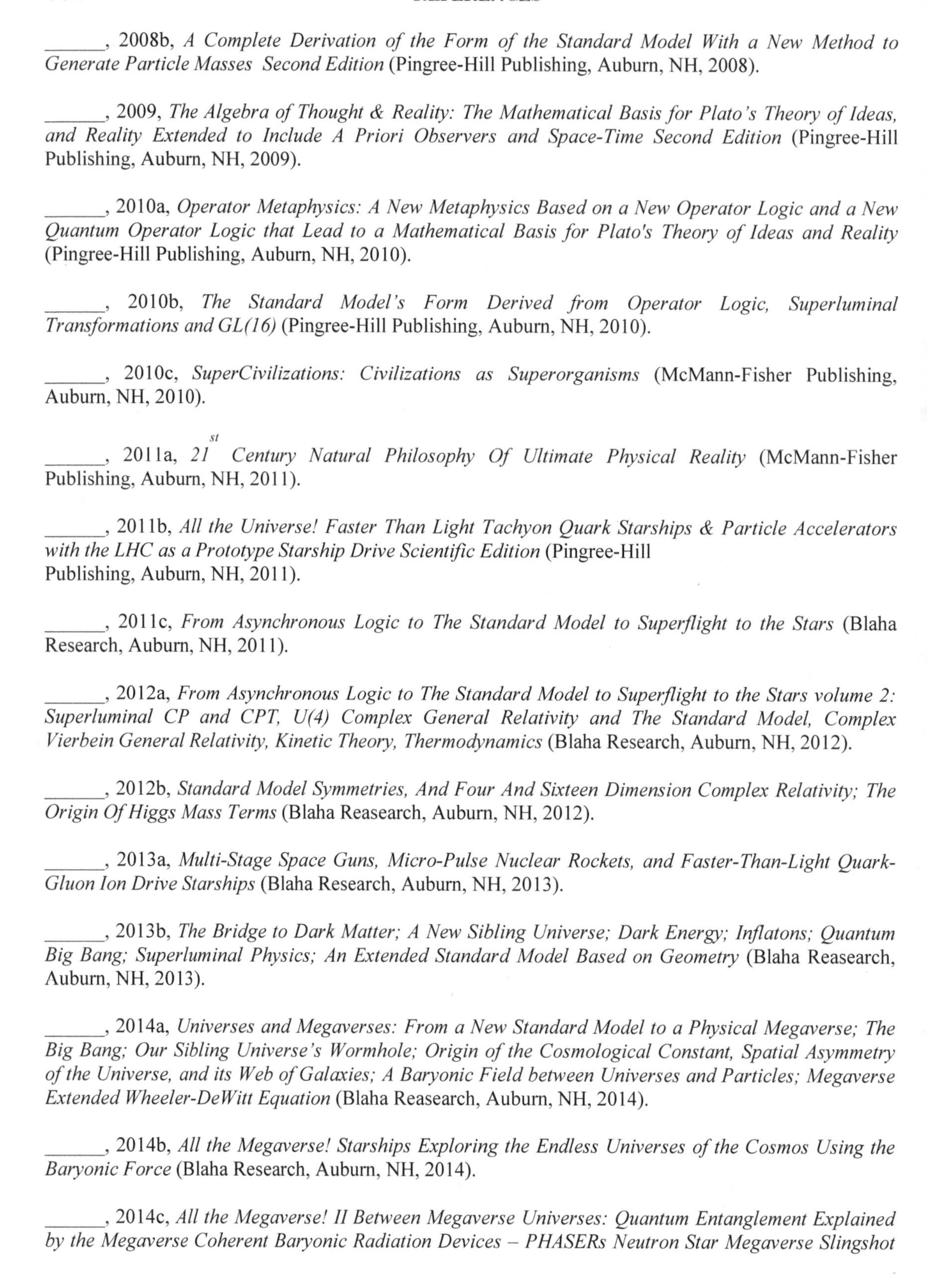

______, 2008b, *A Complete Derivation of the Form of the Standard Model With a New Method to Generate Particle Masses Second Edition* (Pingree-Hill Publishing, Auburn, NH, 2008).

______, 2009, *The Algebra of Thought & Reality: The Mathematical Basis for Plato's Theory of Ideas, and Reality Extended to Include A Priori Observers and Space-Time Second Edition* (Pingree-Hill Publishing, Auburn, NH, 2009).

______, 2010a, *Operator Metaphysics: A New Metaphysics Based on a New Operator Logic and a New Quantum Operator Logic that Lead to a Mathematical Basis for Plato's Theory of Ideas and Reality* (Pingree-Hill Publishing, Auburn, NH, 2010).

______, 2010b, *The Standard Model's Form Derived from Operator Logic, Superluminal Transformations and GL(16)* (Pingree-Hill Publishing, Auburn, NH, 2010).

______, 2010c, *SuperCivilizations: Civilizations as Superorganisms* (McMann-Fisher Publishing, Auburn, NH, 2010).

______, 2011a, *21st Century Natural Philosophy Of Ultimate Physical Reality* (McMann-Fisher Publishing, Auburn, NH, 2011).

______, 2011b, *All the Universe! Faster Than Light Tachyon Quark Starships & Particle Accelerators with the LHC as a Prototype Starship Drive Scientific Edition* (Pingree-Hill Publishing, Auburn, NH, 2011).

______, 2011c, *From Asynchronous Logic to The Standard Model to Superflight to the Stars* (Blaha Research, Auburn, NH, 2011).

______, 2012a, *From Asynchronous Logic to The Standard Model to Superflight to the Stars volume 2: Superluminal CP and CPT, U(4) Complex General Relativity and The Standard Model, Complex Vierbein General Relativity, Kinetic Theory, Thermodynamics* (Blaha Research, Auburn, NH, 2012).

______, 2012b, *Standard Model Symmetries, And Four And Sixteen Dimension Complex Relativity; The Origin Of Higgs Mass Terms* (Blaha Reasearch, Auburn, NH, 2012).

______, 2013a, *Multi-Stage Space Guns, Micro-Pulse Nuclear Rockets, and Faster-Than-Light Quark-Gluon Ion Drive Starships* (Blaha Research, Auburn, NH, 2013).

______, 2013b, *The Bridge to Dark Matter; A New Sibling Universe; Dark Energy; Inflatons; Quantum Big Bang; Superluminal Physics; An Extended Standard Model Based on Geometry* (Blaha Reasearch, Auburn, NH, 2013).

______, 2014a, *Universes and Megaverses: From a New Standard Model to a Physical Megaverse; The Big Bang; Our Sibling Universe's Wormhole; Origin of the Cosmological Constant, Spatial Asymmetry of the Universe, and its Web of Galaxies; A Baryonic Field between Universes and Particles; Megaverse Extended Wheeler-DeWitt Equation* (Blaha Reasearch, Auburn, NH, 2014).

______, 2014b, *All the Megaverse! Starships Exploring the Endless Universes of the Cosmos Using the Baryonic Force* (Blaha Research, Auburn, NH, 2014).

______, 2014c, *All the Megaverse! II Between Megaverse Universes: Quantum Entanglement Explained by the Megaverse Coherent Baryonic Radiation Devices – PHASERs Neutron Star Megaverse Slingshot*

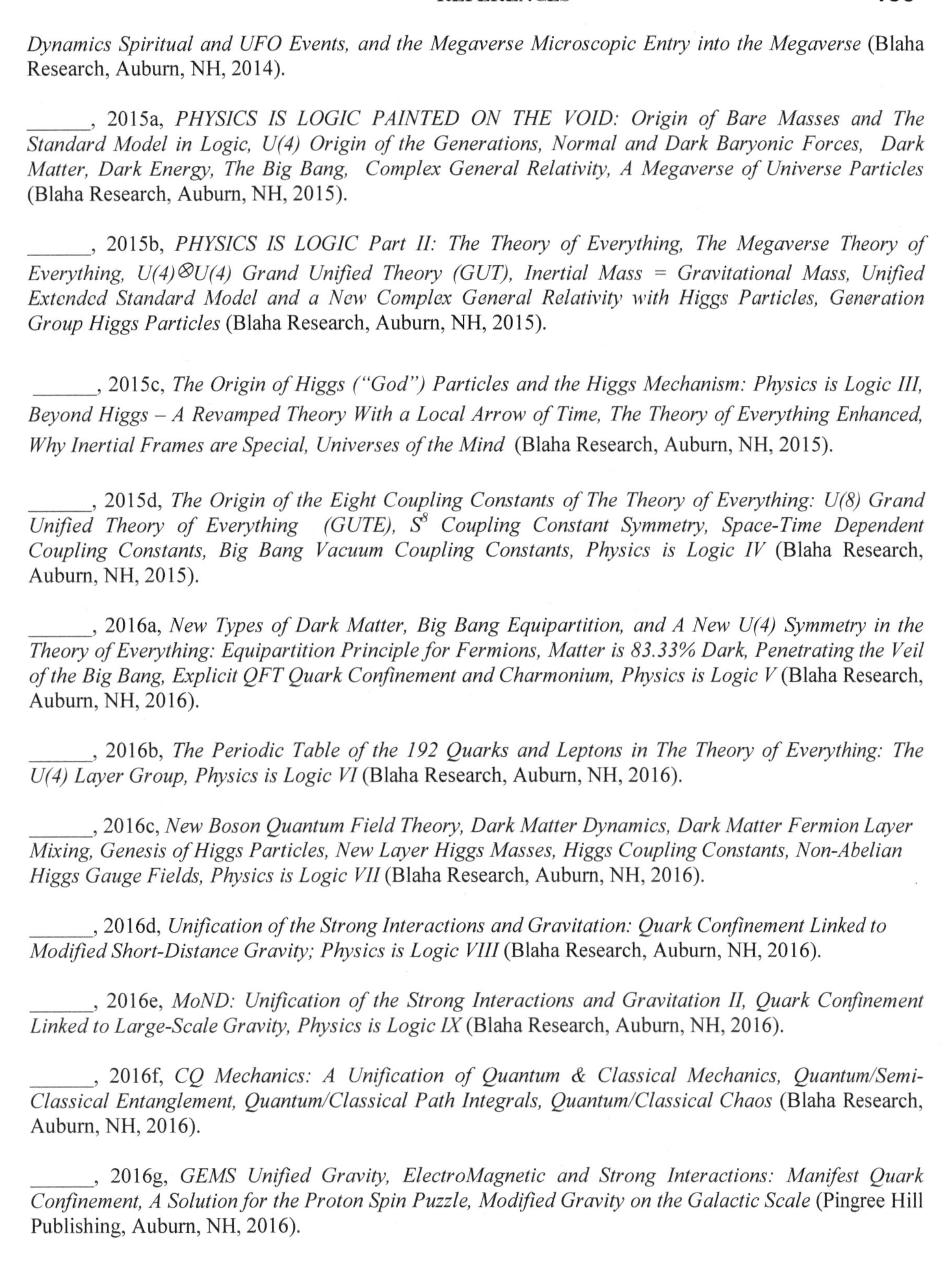

Dynamics Spiritual and UFO Events, and the Megaverse Microscopic Entry into the Megaverse (Blaha Research, Auburn, NH, 2014).

______, 2015a, *PHYSICS IS LOGIC PAINTED ON THE VOID: Origin of Bare Masses and The Standard Model in Logic, U(4) Origin of the Generations, Normal and Dark Baryonic Forces, Dark Matter, Dark Energy, The Big Bang, Complex General Relativity, A Megaverse of Universe Particles* (Blaha Research, Auburn, NH, 2015).

______, 2015b, *PHYSICS IS LOGIC Part II: The Theory of Everything, The Megaverse Theory of Everything,* $U(4)\otimes U(4)$ *Grand Unified Theory (GUT), Inertial Mass = Gravitational Mass, Unified Extended Standard Model and a New Complex General Relativity with Higgs Particles, Generation Group Higgs Particles* (Blaha Research, Auburn, NH, 2015).

______, 2015c, *The Origin of Higgs ("God") Particles and the Higgs Mechanism: Physics is Logic III, Beyond Higgs – A Revamped Theory With a Local Arrow of Time, The Theory of Everything Enhanced, Why Inertial Frames are Special, Universes of the Mind* (Blaha Research, Auburn, NH, 2015).

______, 2015d, *The Origin of the Eight Coupling Constants of The Theory of Everything: U(8) Grand Unified Theory of Everything (GUTE),* S^8 *Coupling Constant Symmetry, Space-Time Dependent Coupling Constants, Big Bang Vacuum Coupling Constants, Physics is Logic IV* (Blaha Research, Auburn, NH, 2015).

______, 2016a, *New Types of Dark Matter, Big Bang Equipartition, and A New U(4) Symmetry in the Theory of Everything: Equipartition Principle for Fermions, Matter is 83.33% Dark, Penetrating the Veil of the Big Bang, Explicit QFT Quark Confinement and Charmonium, Physics is Logic V* (Blaha Research, Auburn, NH, 2016).

______, 2016b, *The Periodic Table of the 192 Quarks and Leptons in The Theory of Everything: The U(4) Layer Group, Physics is Logic VI* (Blaha Research, Auburn, NH, 2016).

______, 2016c, *New Boson Quantum Field Theory, Dark Matter Dynamics, Dark Matter Fermion Layer Mixing, Genesis of Higgs Particles, New Layer Higgs Masses, Higgs Coupling Constants, Non-Abelian Higgs Gauge Fields, Physics is Logic VII* (Blaha Research, Auburn, NH, 2016).

______, 2016d, *Unification of the Strong Interactions and Gravitation: Quark Confinement Linked to Modified Short-Distance Gravity; Physics is Logic VIII* (Blaha Research, Auburn, NH, 2016).

______, 2016e, *MoND: Unification of the Strong Interactions and Gravitation II, Quark Confinement Linked to Large-Scale Gravity, Physics is Logic IX* (Blaha Research, Auburn, NH, 2016).

______, 2016f, *CQ Mechanics: A Unification of Quantum & Classical Mechanics, Quantum/Semi-Classical Entanglement, Quantum/Classical Path Integrals, Quantum/Classical Chaos* (Blaha Research, Auburn, NH, 2016).

______, 2016g, *GEMS Unified Gravity, ElectroMagnetic and Strong Interactions: Manifest Quark Confinement, A Solution for the Proton Spin Puzzle, Modified Gravity on the Galactic Scale* (Pingree Hill Publishing, Auburn, NH, 2016).

______, 2016h, *Unification of the Seven Boson Interactions based on the Riemann-Christoffel Curvature Tensor* (Pingree Hill Publishing, Auburn, NH, 2016).

______, 2017a, *Unification of the Eleven Boson Interactions based on 'Rotations of Interactions'* (Pingree Hill Publishing, Auburn, NH, 2017).

______, 2017b, *The Origin of Fermions and Bosons, and Their Unification* (Pingree Hill Publishing, Auburn, NH, 2017).

______, 2017c, *Megaverse: The Universe of Universes* (Pingree Hill Publishing, Auburn, NH, 2017).

______, 2017d, *SuperSymmetry and the Unified SuperStandard Model* (Pingree Hill Publishing, Auburn, NH, 2017).

______, 2017e, *From Qubits to the Unified SuperStandard Model with Embedded SuperStrings: A Derivation* (Pingree Hill Publishing, Auburn, NH, 2017).

______, 2017f, *The Unified SuperStandard Model in Our Universe and the Megaverse: Quarks, ... ,* (Pingree Hill Publishing, Auburn, NH, 2017).

______, 2018a, *The Unified SuperStandard Model and the Megaverse SECOND EDITION A Deeper Theory based on a New Particle Functional Space that Explicates Quantum Entanglement Spookiness (Volume 1)* (Pingree Hill Publishing, Auburn, NH, 2018).

______, 2018b, *Cosmos Creation: The Unified SuperStandard Model, Volume 2, SECOND EDITION* (Pingree Hill Publishing, Auburn, NH, 2018).

______, 2018c, *God Theory (*Pingree Hill Publishing, Auburn, NH, 2018).

______, 2018d, *Immortal Eye: God Theory: Second Edition* (Pingree Hill Publishing, Auburn, NH, 2018).

______, 2018e, *Unification of God Theory and Unified SuperStandard Model THIRD EDITION* (Pingree Hill Publishing, Auburn, NH, 2018).

______, 2019a, *Calculation of: QED* $\alpha = 1/137$*, and Other Coupling Constants of the Unified SuperStandard Theory* (Pingree Hill Publishing, Auburn, NH, 2019).

______, 2019b, *Coupling Constants of the Unified SuperStandard Theory SECOND EDITION* (Pingree Hill Publishing, Auburn, NH, 2019).

______, 2019c, *New Hybrid Quantum Big_Bang–Megaverse_Driven Universe with a Finite Big Bang and an Increasing Hubble Constant* (Pingree Hill Publishing, Auburn, NH, 2019).

______, 2019d, *The Universe, The Electron and The Vacuum* (Pingree Hill Publishing, Auburn, NH, 2019).

______, 2019e, *Quantum Big Bang – Quantum Vacuum Universes (Particles)* (Pingree Hill Publishing, Auburn, NH, 2019).

______, 2019f, *The Exact QED Calculation of the Fine Structure Constant Implies ALL 4D Universes have the Same Physics/Life Prospects* (Pingree Hill Publishing, Auburn, NH, 2019).

______, 2019g, *Unified SuperStandard Theory and the SuperUniverse Model: The Foundation of Science* (Pingree Hill Publishing, Auburn, NH, 2019).

______, 2020a, *Quaternion Unified SuperStandard Theory (The QUeST) and Megaverse Octonion SuperStandard Theory (MOST)* (Pingree Hill Publishing, Auburn, NH, 2020).

______, 2020b, *United Universes Quaternion Universe - Octonion Megaverse* (Pingree Hill Publishing, Auburn, NH, 2020).

______, 2020c, *Unified SuperStandard Theories for Quaternion Universes & The Octonion Megaverse* (Pingree Hill Publishing, Auburn, NH, 2020).

______, 2020d, *The Essence of Eternity: Quaternion & Octonion SuperStandard Theories* (Pingree Hill Publishing, Auburn, NH, 2020).

______, 2020e, *The Essence of Eternity II* (Pingree Hill Publishing, Auburn, NH, 2020).

______, 2020f, *A Very Conscious Universe* (Pingree Hill Publishing, Auburn, NH, 2020).

______, 2020g, *Hypercomplex Universe* (Pingree Hill Publishing, Auburn, NH, 2020).

______, 2020h, *Beneath the Quaternion Universe* (Pingree Hill Publishing, Auburn, NH, 2020).

______, 2020i, *Why is the Universe Real? From Quaternion & Octonion to Real Coordinates* (Pingree Hill Publishing, Auburn, NH, 2020).

______, 2020j, *The Origin of Universes: of Quaternion Unified SuperStandard Theory (QUeST); and of the Octonion Megaverse (UTMOST)* (Pingree Hill Publishing, Auburn, NH, 2020).

______, 2020k, *The Seven Spaces of Creation: Octonion Cosmology* (Pingree Hill Publishing, Auburn, NH, 2020).

______, 2020l, *From Octonion Cosmology to the Unified SuperStandard Theory of Particles* (Pingree Hill Publishing, Auburn, NH, 2020).

______, 2021a, *Pioneering the Cosmos* (Pingree Hill Publishing, Auburn, NH, 2021).

______, 2021b, *Pioneering the Cosmos II* (Pingree Hill Publishing, Auburn, NH, 2021).

______, 2021c, *Beyond Octonion Cosmology* (Pingree Hill Publishing, Auburn, NH, 2021).

______, 2021d, *Universes are Particles* (Pingree Hill Publishing, Auburn, NH, 2021).

______, 2021e, *Octonion-like dna-based life, Universe expansion is decay, Emerging New Physics* (Pingree Hill Publishing, Auburn, NH, 2021).

______, 2021f, *The Science of Creation New Quantum Field Theory of Spaces* (Pingree Hill Publishing, Auburn, NH, 2021).

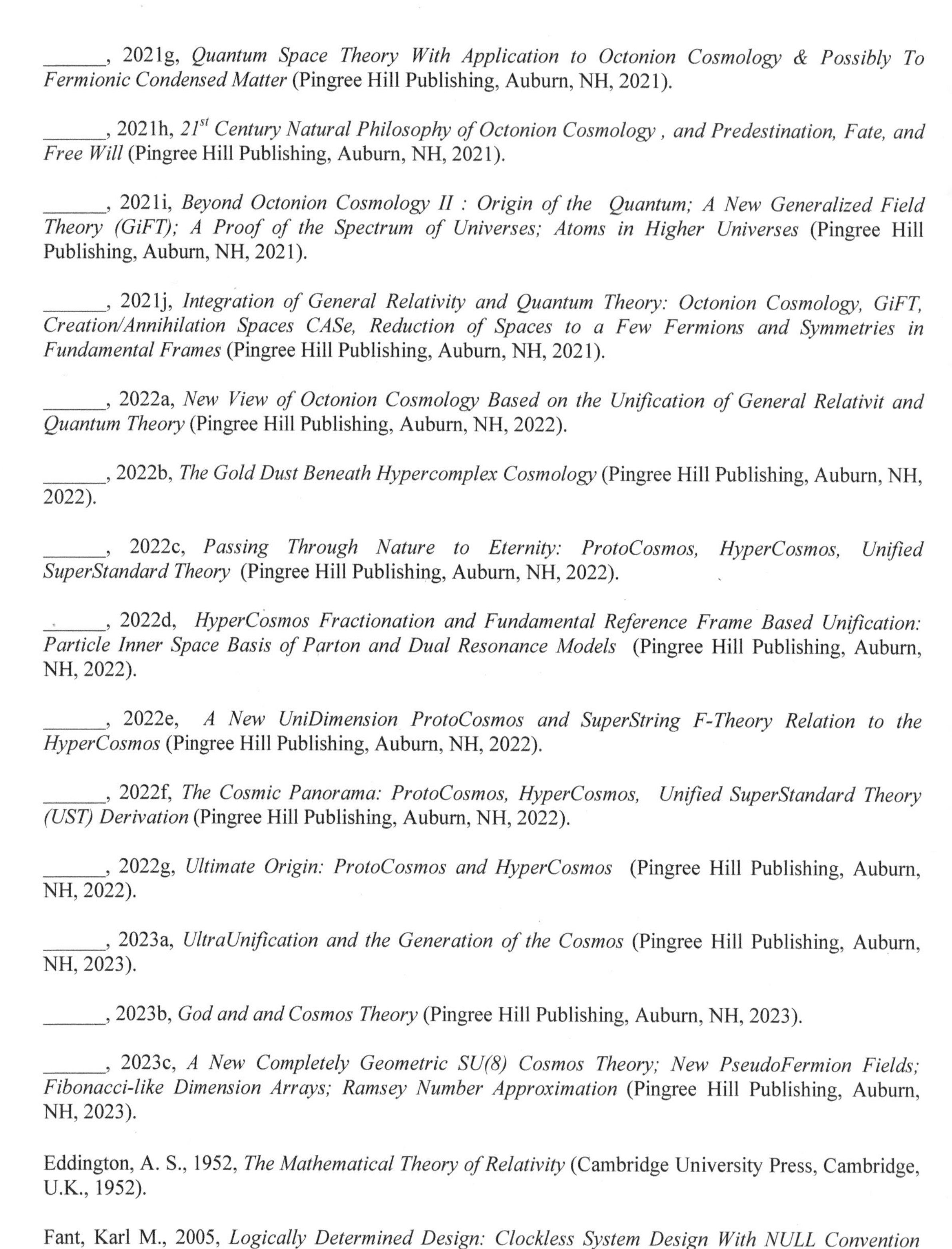

______, 2021g, *Quantum Space Theory With Application to Octonion Cosmology & Possibly To Fermionic Condensed Matter* (Pingree Hill Publishing, Auburn, NH, 2021).

______, 2021h, *21st Century Natural Philosophy of Octonion Cosmology , and Predestination, Fate, and Free Will* (Pingree Hill Publishing, Auburn, NH, 2021).

______, 2021i, *Beyond Octonion Cosmology II : Origin of the Quantum; A New Generalized Field Theory (GiFT); A Proof of the Spectrum of Universes; Atoms in Higher Universes* (Pingree Hill Publishing, Auburn, NH, 2021).

______, 2021j, *Integration of General Relativity and Quantum Theory: Octonion Cosmology, GiFT, Creation/Annihilation Spaces CASe, Reduction of Spaces to a Few Fermions and Symmetries in Fundamental Frames* (Pingree Hill Publishing, Auburn, NH, 2021).

______, 2022a, *New View of Octonion Cosmology Based on the Unification of General Relativit and Quantum Theory* (Pingree Hill Publishing, Auburn, NH, 2022).

______, 2022b, *The Gold Dust Beneath Hypercomplex Cosmology* (Pingree Hill Publishing, Auburn, NH, 2022).

______, 2022c, *Passing Through Nature to Eternity: ProtoCosmos, HyperCosmos, Unified SuperStandard Theory* (Pingree Hill Publishing, Auburn, NH, 2022).

______, 2022d, *HyperCosmos Fractionation and Fundamental Reference Frame Based Unification: Particle Inner Space Basis of Parton and Dual Resonance Models* (Pingree Hill Publishing, Auburn, NH, 2022).

______, 2022e, *A New UniDimension ProtoCosmos and SuperString F-Theory Relation to the HyperCosmos* (Pingree Hill Publishing, Auburn, NH, 2022).

______, 2022f, *The Cosmic Panorama: ProtoCosmos, HyperCosmos, Unified SuperStandard Theory (UST) Derivation* (Pingree Hill Publishing, Auburn, NH, 2022).

______, 2022g, *Ultimate Origin: ProtoCosmos and HyperCosmos* (Pingree Hill Publishing, Auburn, NH, 2022).

______, 2023a, *UltraUnification and the Generation of the Cosmos* (Pingree Hill Publishing, Auburn, NH, 2023).

______, 2023b, *God and and Cosmos Theory* (Pingree Hill Publishing, Auburn, NH, 2023).

______, 2023c, *A New Completely Geometric SU(8) Cosmos Theory; New PseudoFermion Fields; Fibonacci-like Dimension Arrays; Ramsey Number Approximation* (Pingree Hill Publishing, Auburn, NH, 2023).

Eddington, A. S., 1952, *The Mathematical Theory of Relativity* (Cambridge University Press, Cambridge, U.K., 1952).

Fant, Karl M., 2005, *Logically Determined Design: Clockless System Design With NULL Convention Logic* (John Wiley and Sons, Hoboken, NJ, 2005).

Feinberg, G. and Shapiro, R., 1980, *Life Beyond Earth: The Intelligent Earthlings Guide to Life in the Universe* (William Morrow and Company, New York, 1980).

Gelfand, I. M., Fomin, S. V., Silverman, R. A. (tr), 2000, *Calculus of Variations* (Dover Publications, Mineola, NY, 2000).

Giaquinta, M., Modica, G., Souchek, J., 1998, *Cartesian Coordinates in the Calculus of Variations* Volumes I and II (Springer-Verlag, New York, 1998).

Giaquinta, M., Hildebrandt, S., 1996, *Calculus of Variations* Volumes I and II (Springer-Verlag, New York, 1996).

Gradshteyn, I. S. and Ryzhik, I. M., 1965, *Table of Integrals, Series, and Products* (Academic Press, New York, 1965).

Heitler, W., 1954, *The Quantum Theory of Radiation* (Claendon Press, Oxford, UK, 1954).

Huang, Kerson, 1992, *Quarks, Leptons & Gauge Fields 2nd Edition* (World Scientific Publishing Company, Singapore, 1992).

Jost, J., Li-Jost, X., 1998, *Calculus of Variations* (Cambridge University Press, New York, 1998).

Kaku, Michio, 1993, *Quantum Field Theory*, (Oxford University Press, New York, 1993).

Kirk, G. S. and Raven, J. E., 1962, *The Presocratic Philosophers* (Cambridge University Press, New York, 1962).

Landau, L. D. and Lifshitz, E. M., 1987, *Fluid Mechanics 2nd Edition*, (Pergamon Press, Elmsford, NY, 1987).

Misner, C. W., Thorne, K. S., and Wheeler, J. A., 1973, *Gravitation* (W. H. Freeman, New York, 1973).

Rescher, N., 1967, *The Philosophy of Leibniz* (Prentice-Hall, Englewood Cliffs, NJ, 1967).

Rieffel, Eleanor and Polak, Wolfgang, 2014, *Quantum Computing* (MIT Press, Cambridge, MA, 2014).

Riesz, Frigyes and Sz.-Nagy, Béla, 1990, *Functional Analysis* (Dover Publications, New York, 1990).

Sagan, H., 1993, *Introduction to the Calculus of Variations* (Dover Publications, Mineola, NY, 1993).

Sakurai, J. J., 1964, *Invariance Principles and Elementary Particles* (Princeton University Press, Princeton, NJ, 1964).

Weinberg, S., 1972, *Gravitation and Cosmology* (John Wiley and Sons, New York, 1972).

Weinberg, S., 1995, *The Quantum Theory of Fields Volume I* (Cambridge University Press, New York, 1995).

INDEX

About The Author

Stephen Blaha is a well-known Physicist and Man of Letters with interests in Science, Society and civilization, the Arts, and Technology. He had an Alfred P. Sloan Foundation scholarship in college. He received his Ph.D. in Physics from Rockefeller University. He has served on the faculties of several major universities. He was also a Member of the Technical Staff at Bell Laboratories, a manager at the Boston Globe Newspaper, a Director at Wang Laboratories, and President of Blaha Software Inc. and of Janus Associates Inc. (NH).

Among other achievements he was a co-discoverer of the "r potential" for heavy quark binding developing the first (and still the only demonstrable) non-Aeolian gauge theory with an "r" potential; first suggested the existence of topological structures in superfluid He-3; first proposed Yang-Mills theories would appear in condensed matter phenomena with non-scalar order parameters; first developed a grammar-based formalism for quantum computers and applied it to elementary particle theories; first developed a new form of quantum field theory without divergences (thus solving a major 60 year old problem that enabled a unified theory of the Standard Model and Quantum Gravity without divergences to be developed); first developed a formulation of complex General Relativity based on analytic continuation from real space-time; first developed a generalized non-homogeneous Robertson-Walker metric that enabled a quantum theory of the Big Bang to be developed without singularities at t = 0; first generalized Cauchy's theorem and Gauss' theorem to complex, curved multi-dimensional spaces; received Honorable Mention in the Gravity Research Foundation Essay Competition in 1978; first developed a physically acceptable theory of faster-than-light particles; first derived a composition of extremums method in the Calculus of Variations; first quantitatively suggested that inflationary periods in the history of the universe were not needed; first proved Gödel's Theorem implies Nature must be quantum; provided a new alternative to the Higgs Mechanism, and Higgs particles, to generate masses; first showed how to resolve logical paradoxes including Gödel's Undecidability Theorem by developing Operator Logic and Quantum Operator Logic; first developed a quantitative harmonic oscillator-like model of the life cycle, and interactions, of civilizations; first showed how equations describing superorganisms also apply to civilizations. A recent book shows his theory applies successfully to the past 14 years of history and to *new* archaeological data on Andean and Mayan civilizations as well as Early Anatolian and Egyptian civilizations.

He first developed an axiomatic derivation of the form of The Standard Model from geometry – space-time properties – The Unified SuperStandard Model. It unifies all the known forces of Nature. It also has a Dark Matter sector that includes a Dark ElectroWeak sector with Dark doublets and Dark gauge interactions. It uses quantum coordinates to remove infinities that crop up in most interacting quantum field theories and additionally to remove the infinities that appear in the Big Bang and generate inflationary growth of the universe. It shows gravity has a MOND-like form without sacrificing Newton's Laws. It relates the interactions of the MOND-like sector of gravity with the r-potential of Quark Confinement. The axioms of the theory lead to the question of their origin. We suggest in the preceding edition of this book it can be attributed to an entity with God-like properties. We explore these properties in "God Theory" and show they predict that the Cosmos exists forever although individual universes (or incarnations of our universe) "come and go." Several other important results emerge from God Theory such a functionally triune God. The Unified SuperStandard Theory has many other important parts described in the Current Edition of *The Unified SuperStandard Theory* and expanded in subsequent volumes.

Blaha has had a major impact on a succession of elementary particle theories: his Ph.D. thesis (1970), and papers, showed that quantum field theory calculations to all orders in ladder approximations could not give scaling deep inelastic electron-nucleon scattering. He later showed the eigenvalue equation for the fine structure constant α in Johnson-Baker-Willey QED had a zero at $\alpha = 1$ not 1/137 by solving the Schwinger-Dyson equations to all orders in an approximation that agreed with exact results to 4th order in α thus ending interest in this theory. In 1979 at Prof. Ken Johnson's (MIT) suggestion he calculated the proton-neutron mass difference in the MIT bag model and found the result had the wrong sign reducing interest in the bag model. These results all appear in Physical Review papers. In the 2000's he repeatedly pointed out the shortcomings of SuperString theory and showed that The Standard Model's form could be derived from space-time geometry by an extension of Lorentz transformations to faster than light transformations. This deeper space-time basis greatly increases the possibility that it is part of THE fundamental theory. Recently, Blaha showed that the Weak interactions differed significantly from the Strong, electromagnetic and gravitation interactions in important respects while these interactions had similar features, and suggested that ElectroWeak theory, which is essentially a glued union of the Weak interactions and Electromagnetism, possibly modulo unknown Higgs particle features, be replaced by a unified theory of the other interactions combined with a stand-alone Weak interaction theory. Blaha also showed that, if Charmonium calculations are taken seriously, the Strong interaction coupling constant is only a factor of five larger than the electromagnetic coupling constant, and thus Strong interaction perturbation theory would make sense and yield physically meaningful results.

In graduate school (1965-71) he wrote substantial and significant papers in elementary particles and group theory: The Inelastic E- P Structure Functions in a Gluon Model. Phys. Lett. B40:501-502,1972; Deep-Inelastic E-P Structure Functions In A Ladder Model With Spin 1/2 Nucleons, Phys.Rev. D3:510-523,1971; Continuum Contributions To The Pion Radius, Phys. Rev. 178:2167-2169,1969; Character Analysis of U(N) and SU(N), J. Math. Phys. 10, 2156 (1969); and The Calculation of the Irreducible Characters of the Symmetric Group in Terms of the Compound Characters, (Published as Blaha's Lemma in D. E. Knuth's book: *The Art of Computer Programming Vols. 1 – 4*).

In the early 1980's Blaha was also a pioneer in the development of UNIX for financial, scientific and Internet applications: benchmarked UNIX versions showing that block size was critical for UNIX performance, developing financial modeling software, starting database benchmarking comparison studies, developing Internet-like UNIX networking (1982) and developing a hybrid shell programming technique (1982) that was a precursor to the PERL programming language. He was also the manager of the AT&T ten-year future products development database. His work helped lead to commercial UNIX on computers such as Sun Micros, IBM AIX minis, and Apple computers.
In the 1980's he pioneered the development of PC Desktop Publishing on laser printers and was nominated for three "Awards for Technical Excellence" in 1987 by PC Magazine for PC software products that he designed and developed.
Recently he has developed a theory of Megaverses – actual universes of which our universe is one – with quantum particle-like properties based on the Wheeler-DeWitt equation of Quantum Gravity. He has developed a theory of a baryonic force, which had been conjectured many years ago, and estimated the strength of the force based on discrepancies in measurements of the gravitational constant G. This force, operative in D-dimensional space, can be used to escape from our universe in "uniships" which are the equivalent of the faster-than-light starships proposed in the author's earlier books. Thus travel to other universes, as well as to other stars is possible.
Blaha also considered the complexified Wheeler-DeWitt equation and showed that its limitation to real-valued coordinates and metrics generated a Cosmological Constant in the Einstein equations.
The author has also recently written a series of books on the serious problems of the United States and their solution as well as a book on the decline of Mankind that will follow from current social and genetic trends in Mankind.

In the past twenty years Dr. Blaha has written over 80 books on a wide range of topics. Some recent major works are: *From Asynchronous Logic to The Standard Model to Superflight to the Stars, All the Universe!, SuperCivilizations: Civilizations as Superorganisms, America's Future: an Islamic Surge, ISIS, al Qaeda, World Epidemics, Ukraine, Russia-China Pact, US Leadership Crisis, The Rises and Falls of Man – Destiny – 3000 AD: New Support for a Superorganism MACRO-THEORY of CIVILIZATIONS From CURRENT WORLD TRENDS and NEW Peruvian, Pre-Mayan, Mayan, Anatolian, and Early Egyptian Data, with a Projection to 3000 AD,* and *Mankind in Decline: Genetic Disasters, Human-Animal Hybrids, Overpopulation, Pollution, Global Warming, Food and Water Shortages, Desertification, Poverty, Rising Violence, Genocide, Epidemics, Wars, Leadership Failure.*

He has taught approximately 4,000 students in undergraduate, graduate, and postgraduate corporate education courses primarily in major universities, and large companies and government agencies.

He developed a quantum theory, The Unified SuperStandard Theory (UST), which describes elementary particles in detail without the difficulties of conventional quantum field theory. He found that the internal symmetries of this theory could be exactly derived from an octonion theory called QUeST. He further found that another octonion theory (UTMOST) describes the Megaverse. It can hold QUeST universes such as our own universe. It has an internal symmetry structure which is a superset of the QUeST internal symmetries.

Recently he developed Octonion Cosmology. He replaced it with HyperCosmos theory, which has significantly better features. He developed a fractionalization process for dimensions, particles and symmetry groups. He also described transformation that reduced particles and dimensions to a far more compact form. He also developed a precursor theory ProtoCosmos that leads to the HyperCosmos.

The author showed that space-time and Internal Symmetries can be unified in any of the ten HyperCosmos spaces in their associated HyperUnification spaces. The combined set of HyperUnification spaces enable all HyperCosmos dimensions to be obtained by a General Relativistic transformation from one primordial dimension in the 42 space-time dimension unified HyperUnification space.

At present the author has a Cosmos Theory that incorporates ProtoCosmos Theory, HyperCosmos Theory, Limos Theory, Second Kind HyperCosmos Theory and HyperUnification Spaces. He has introduced PseudoFermion wave functions and theory, He has related Cosmos Theory to Regge trajectories of spaces, parton theory, Veneziano amplitudes, Fibonacci numbers and Ramsey numbers. He has calculated an approximation to the difficult to calculate R(n,n) Ramsey numbers.

Ingram Content Group UK Ltd.
Milton Keynes UK
UKHW052037210723
425569UK00003B/64

9 781735 679563